D0435067

Life through Time and Space

LIFE *through* TIME *and* SPACE

WALLACE ARTHUR

Harvard University Press
Cambridge, Massachusetts & London, England
2017

Printed in the United States of America

First Printing

Library of Congress Cataloging-in-Publication Data

Names: Arthur, Wallace, author. | Arthur, Stephen, illustrator.
Title: Life through time and space / Wallace Arthur;
Illustrations by Stephen Arthur.
Description: Cambridge, Massachusetts : Harvard University Press, 2017. |
Includes bibliographical references and index.
Identifiers: LCCN 2016058797 | ISBN 9780674975866 (cloth)
Subjects: LCSH: Life—Origin. | Evolution. | Human evolution. | Life on other planets. | Exobiology. | Extraterrestrial beings. | Developmental biology.
Classification: LCC QH325 .A724 2017 | DDC 576.8/3—dc23
LC record available at https://lccn.loc.gov/2016058797

The known is finite, the unknown infinite; intellectually we stand
on an islet in the midst of an illimitable ocean of inexplicability.
Our business in every generation is to reclaim a little more land,
to add something to the extent and the solidity of our possessions.

—THOMAS HENRY HUXLEY, 1887

Contents

Illustrations by Stephen Arthur

Preface

I've spent most of my life as a biologist; and in the last few years I've spent a lot of time studying astronomy. This book is a celebration of these sciences and of the growing relationship between them—a celebratory story told in plain language, not technical jargon. It's about our origins, our fates, our place in the universe, and the likelihood of intelligent alien life, looked at from both biological and astronomical points of view. I dedicate the book to all those who help to promote understanding rather than indoctrination, and in particular that great truth-seeker Thomas Henry Huxley. *Understanding* is one of the noblest goals of humanity, and one of the keys to our survival and progress.

The book is written as a series of seven triplets of chapters. Within each triplet, the first chapter is predominantly astronomical or astrobiological in flavor, the second is evolutionary, and the third is embryological. However, each chapter has arms that reach out into one or both of the other two domains. The connections between triplets, and between chapters within triplets, might at first seem cryptic, but they have a curious logic. I can best illustrate the nature of these connections by using the first triplet as an example.

In Chapter 1 we consider hypothetical (but maybe also real) inhabitants of the Andromeda galaxy, an object that can be seen in the nighttime sky by anyone with reasonably good eyesight. We use these extraterrestrial creatures as an entry point to the possibility of alien life in general. We imagine them looking toward the Earth with telescopes so advanced that they can actually see not just our planet but individual people wandering over its surface.

In Chapter 2 we meet some of the people they see. But these would not be us. Rather, since light from Andromeda takes 2.5 million years to reach us, and the same span of time to travel in the other direction, they would see protopeople of the distant past, belonging to an early species of *Homo,* characterized by a brain that's about half the size of our own.

In Chapter 3 we acknowledge that characterizing a species of human or protohuman as having a brain of a specific size is an oversimplification. As individuals, our brains are at first nonexistent, then small, then large. Early human embryos lack not only brains but any nerve cells at all. So, we contemplate the form of these early embryos, and the question of how they go about producing the beginnings of their nervous systems.

The other connections between chapters work in similar ways, so I won't waste words elucidating them further here. Their individual details will reveal themselves soon enough. The overall pattern of their linkage is designed to take you on a fascinating journey through embryological and evolutionary time, terrestrial and interstellar space.

Life through Time and Space

FROM STARS TO EMBRYOS

Chapter One

Galaxy Gazing

The Big W in the Sky

On a cloudless night, step outside and gaze up at the sky. What do you see? The short answer to this question is suns—or stars, which is a different name for the same thing. There are lots of them; just how many you see depends on where you live. If you're in a big city, you'll perhaps see only tens of them. If you're deep in the countryside, where the level of light pollution is low, you'll be able to see hundreds or, if you use binoculars, thousands. Not quite all of the bright objects you see in the sky will be stars. The exceptions will probably be a couple of planets, the lights of a few planes, and perhaps an orbiting satellite. What I want to direct your attention to, though, is a strange fuzzy blob which is none of these things. It's a galaxy, and one that almost certainly harbors intelligent life. When you look up at it, there is someone, or something, looking back at you.

The object I'm thinking of is the great galaxy of Andromeda. To find it, you can use a group of five bright stars that form the shape of a W. This is the constellation Cassiopeia, named after a vain queen in Greek mythology. It's one of the most conspicuous in the nighttime sky, if you live in the Northern Hemisphere. It can be seen year-round, even from quite light-polluted localities. Once you know that there's a big W in the sky, it's quite easy to find. Of course, a W is really two Vs stuck together. To locate the Andromeda galaxy, you use the right-hand V of this particular W as an arrowhead.

Here's what you do. Look at how deep the V is. Then project the angle of your view down about three times that depth, in the direction of the arrowhead and very slightly skewed to the right. There you will find the Andromeda galaxy.

How easy is this object to see with the naked eye? This depends on just three things. The first is your eyesight. If you have good sight, you should see it. If, like me, you have only average sight, you might see it and you might not—but with a pair of binoculars you'll be fine. The second is the level of light pollution. If you live in a large city, you might have to go out into the surrounding countryside to be able to see the galaxy. The third is the time of year. Although the big W is visible in all seasons, when you project the direction of your gaze the appropriate distance in the direction of the arrowhead, there is a time of year when this will take you to a point close to, or below, the horizon. For many of us, this will be just a short period in the spring; the exact duration depends on where you live.

Meet the Andromedans

Once you see the galaxy, consider the following. It looks like a little fuzzy blob, but it consists of about 500 billion suns (or stars). Orbiting most of those suns are planets, some of them much like our own Earth. On many of those planets, there are life-forms. This may at first seem like an overly strong statement: why "many" rather than "a few"? And why no "probably"? We'll soon see the answers to these questions; for the moment, please trust me that the sums work out in such a way that the likelihood of there being no life at all in the Andromeda galaxy is negligible.

Let's imagine two humanoid Andromedan scientists looking exactly in our direction, with a hypothetical telescope so powerful that they can see not just our planet but also individual humans walking across its surface. Even if such a device were possible, when these Andromedans look directly at us they do not see us. Why not?

Space and Time

To answer this question, we need to think a bit about space and time. The Andromeda galaxy is very close to us, as galaxies go. Admittedly it's many trillions of kilometers, or miles, away, but it's within our "local group" of galaxies (I love that phrase)—the ones that are really, really close to us, compared with all the others. Since neither kilometers nor miles work well for us when it comes to comprehending the vast distances of intergalactic space, we use other units, one of which is the light-year. We'll have to get thoroughly on top of this unit to understand exactly how far away Andromeda and other galaxies are from us. It has to become something that's as familiar to us as a meter or an inch.

The first thing to be clear about is that, despite its name, a light-year is a measure of distance, not time. You may already know this, in which case you'll probably also know that it's the distance light travels through space in a year. But how far is that? We can easily work it out. I was taught, as a child, that light travels at 186,000 miles per second. And so it does. But if you were taught using metric rather than U.S. customary or imperial units, then you'll have been told that light travels at 300,000 kilometers per second—which is the same thing, though a suspiciously round figure. If we know how far light travels in a second, it's easy to calculate how far it will travel in a longer period of time, like a year. To save you doing the sums, here's the answer: very roughly 10 trillion kilometers, or 6 trillion miles (these figures are not suspiciously neat; I've just rounded them to the nearest trillion).

Using this astro-friendly unit, how far away is the Andromeda galaxy? The answer: approximately 2.5 million light-years. Close enough to be "local" in astronomic terms (many galaxies are *billions* of light-years from us), but rather a long way from us in any other terms.

Now let's move from distances in space to distances in time. Actually, this is very straightforward, given that we're starting with the distance unit that we call a light-year. The time that light takes to reach us from Andromeda is, by definition, 2.5 million years. So when we look at the galaxy from Earth, we see it as it was 2.5 million years ago, when the light we're seeing right now was originally emitted from the galaxy and began traveling toward us.

Watching Our Ancestors

This looking back in time works both ways. So if those Andromedan scientists were looking in our precise direction last night, they won't have seen us. But they may have seen some protohumans, perhaps belonging to the species *Homo habilis* (literally, "handy man"; more information on these creatures will follow shortly).

I'm quite convinced that my hypothetical Andromedan scientists are real. Here's why. Observations made over the last couple of decades on stars / suns within our own galaxy—the Milky Way—show that many suns, not just our own, have planets. Not only that, but suns with multiple rather than single planets are common, and may well be the norm. Solar systems with one, two, three, four, five, six, seven, and eight planets are all known—but with the proviso that in each case the figure I quote is a minimum because other planets in the appropriate system may yet remain undiscovered. It seems likely that systems with more than our own eight planets exist too, and will be found soon.

Planets of Life

So there are lots of planets. But how many are Earth-like? The best guess at the time of writing is about 1 in 200, though this figure may well have changed a bit by the time you're reading this chapter, given that a typical book has a gestation period of about a year and the current rate of planet discovery is remarkably high.

This figure of 1 in 200 is reached as follows. We now know of about 4,000 confirmed exoplanets—the name given to planets orbiting suns other than our own. Of these, about 20 are Earth-like, though of course that leaves open the question of exactly *how* Earth-like. If this is a fair sample of our galaxy overall (the Milky Way is a bit smaller than the Andromeda galaxy), then, when we know more, the numbers 20 and 4,000 will simply be scaled up and the fraction of Earth-like planets will remain about the same. Assuming that the Milky Way and Andromeda are broadly similar in their composition, which seems likely, the fraction of planets that are Earth-like there will be approximately the same as it is here.

The route from the Andromeda galaxy to the likelihood of Andromedan life-forms works something like this. We'll guesstimate the number of planets in Andromeda as being the same as the number of stars—about 500 billion. That's probably an underestimate, but no matter; in fact, it's sensible to err on the cautious side when trying to estimate the likelihood of life. Now we can guesstimate the number of Earth-like planets as being 1/200 of this huge number, which works out to 2.5 billion. We'll be pessimistic about the fraction of these that embark on an evolutionary process that produces life—say, just 1 in 100, which is probably another underestimate.

This gives us 25 million planets with life. On what fraction of these has evolution produced *intelligent* life? Let's go with our 1 in 100 fraction again, so we're now down to 250,000 planets. So our best guess is that intelligent life exists on about 250,000 planets within the Andromeda galaxy. Well, actually, no, because there's another factor to be taken into account: how long does intelligent life tend to last once it has evolved? We can't yet answer this question. Nevertheless, the idea that intelligent life is almost always doomed to an early grave would have to be taken to extreme lengths to reduce the guesstimated number of Andromedan civilizations from a quarter of a million to zero.

Anyhow, that's as far as I want to take the guesstimates. The American astronomer Frank Drake, one of the pioneers of SETI (search for extraterrestrial intelligence), went a bit further. He produced an equation for, effectively, the likelihood of intelligent life—not in Andromeda but in our own Milky Way galaxy. However, I don't think we need equations to grasp the general picture, and such pictures are what this book is all about.

Relatedness across the Sky

Now that we've established a high likelihood of the existence of intelligent Andromedans, let's ask how they are related to us. In one sense, the answer is not at all. But let's expand on that a little.

To think about our relatedness to any creature here on Earth, it's helpful to use the letter Y. Time runs upward through the letter, and two processes that separate from each other over time do so at the point where the stem of the letter splits. Take the relationship between humans and chimps. In this case the point of separation was about 7 million years ago. If, instead, we inquire about the relatedness of humans and dogs, the point of separation, or divergence, was closer to 70 million years ago than 7 million.

By choosing ever more distantly related creatures, we can reach ever earlier divergence times. The point at which the lineage leading to the genus *Homo* (humans) diverged from the lineage leading to

Spongia (a genus of sponges, unsurprisingly) was probably around 700 million years ago—up by another order of magnitude. But there is no creature to which our relatedness is such that our lineages diverged from each other another order of magnitude longer ago than this, because 7 *billion* years ago the Earth had not yet been born.

Now, back to the Andromedans. Is it futile to seek a Y that connects them and us? Is a divergence time for their lineage and ours a meaningful concept? Probably yes and no, respectively. But let's not dismiss the idea of relatedness without digging a little deeper. Some people, including the Swedish chemist Svante Arrhenius and the British astronomer Sir Fred Hoyle, have speculated about the possibility of life on Earth having originated from the arrival here of a spore or other hardy life-unit from space—the idea being that life originated somewhere else (where?) and that our lineage is ultimately related to lineages that may still exist on some other planet. If interstellar migration of life-forms is possible (as a biologist, I doubt it), then there could indeed be a Y that connects the human lineage here with a humanoid lineage elsewhere.

However unlikely the interstellar migration of life-forms, the likelihood of *intergalactic* migration is even lower. What kind of spore could survive more than 2 million years of space travel at the speed of light, or even longer at a lower speed? None that we know of, for sure. So in the end it's likely that no living stem connects us with Andromedan life-forms. But what about a non-living stem? In the conventional evolutionary biology of Earth-bound creatures, such a possibility is (quite rightly) not considered. But let's think outside the proverbial box.

What I'm getting at here is the question of whether there was some ancient material—perhaps gas or dust—from which both the Milky Way and Andromeda, and hence all their constituent creatures, formed. Strangely, there is not yet a consensus about this issue: we remain somewhat in the dark about how galaxies were born. We'll come back to that particular type of origin later. For now, let's just say that if there was a common cloud from which we and the Andromedans came, it existed many billions of years ago.

Life on the Wing?

You may have noticed something interesting that snuck in untrumpeted in the previous paragraph. This was the idea of all the "constituent creatures" of the Milky Way—the implication being, of course, that there is alien life much closer to us than Andromeda. If that's true (it probably is, but we don't know for sure), why have I started out by asking you to consider the possibility of life so far away? The answer is that I want you to be able to look at one specific thing in the sky where we're pretty sure life exists. When that thing is a fuzzy blob that contains billions of suns, we can indeed be pretty sure. However, when it's a single star, the chances of there being life on one of its orbiting planets are actually quite low.

You'll recall that in the recipe for finding the Andromeda galaxy the first step was to locate the big W in the sky that we call Cassiopeia. The second was to consider the W as being made up of two Vs, the right-hand one of which we used as an arrowhead. The three stars of that arrowhead, from the right-hand edge inward, are Caph, Schedar (the spelling is somewhat variable), and Navi. Perhaps we didn't need to use these as an arrowhead to point to something else—perhaps these suns/stars may themselves have orbiting planets with life. So far, there is no evidence to suggest that they do. However, if we journey to another constellation in the northern sky, Cygnus the swan, there is a star called Kepler 186 that has a planet (186f) that is rather Earth-like and may well host life. This solar system is in the area of the swan's right wing. And there are many others in the same general direction.

The Excitement of Science

Now here's an important issue for what we call "popular science." The main aim of this difficult endeavor is to spread the findings, and indeed the *excitement,* of science, with a minimum of turgid detail. To *do* science, details are crucial. But to *learn* about science's big picture of things, they're not. Or, to be a bit more accurate, they can be minimized. And that's what I've been trying to do so far in this book, and will continue to do throughout, following in the tradition of others who have written in this genre.

But wait a minute: have I succeeded up to this point? Maybe not. Here's a list of the astronomical terms I've mentioned so far: *Cassiopeia, Andromeda, Milky Way, galaxy, light-year, exoplanet, Caph, Schedar, Navi, Cygnus, Kepler 186* (sun), *Kepler 186f* (planet).

That's already a dozen potentially new names. If you're an astronomer, probably none of them will actually be new. But for most people some will be new, and for some people most will be new. How can anyone commit to memory a list of new names without getting bored and losing sight of the big picture and the excitement of scientific discovery? It's vital to provide an answer to this question, for otherwise the mysteries of the universe will be eclipsed by detail, jargon, names. Any author guilty of achieving that appalling eclipse (and there are many) should be ashamed. I will try very hard not to fall into the jargon trap, though 12 potentially new names in fewer than that number of pages does not seem an auspicious start.

But there's a solution to this problem: replacing many individual names with a single framework on which to hang them. For the names we've encountered so far, here's such a framework.

Close, Middling, and Far

There are three domains of space: close, middling, and far. *Close* contains only our own solar system—the Sun, the Earth, the other familiar planets (Mercury, Venus, and so on), the Moon, and a motley collection of other things (asteroids, dwarf planets, comets). From the perspective of another star, such as Caph, this whole collection of stuff can be thought of as just a pinpoint in space. Indeed, that's exactly what it would look like from Caph—it would appear as a fairly ordinary "star" that, if magically zoomed in upon, would reveal all this extraordinary detail including, ultimately, humans.

Middling contains all the other stars of our Milky Way galaxy. Take the arrowhead of Caph, Schedar, and Navi, for example. Although the arrowhead and the W of which it is a part seem like flat entities in the night sky (as do constellations in general, because we can't really detect the third dimension of celestial depth), they're very far from flat indeed. Caph is about 50 light-years away. Schedar, at about 200 light-years, is roughly four times as distant. And Navi is approximately three times farther again, at about 600 light-years. However, in one important sense, these distances are all the same—that is to say, they're all middling.

To see how middling differs from close, consider this. The full span of our solar system (comets and other oddballs aside) from the Sun to the average orbital distance of the farthest-out planet, Neptune, is only a tiny fraction of a light-year—less than a thousandth, in fact. We don't even use the light-year as a unit of measurement at this spatial scale. But the closest star to us—in other words, the closest sun apart from ours—is more than four light-years away. So the closest star is more than 4,000 times as far away from us as is the farthest planet of our solar system. Truly, the close and the middling are different realms of space.

The same is true of the middling and the far. The nearest large galaxy to us, Andromeda, is more than 20 times farther away than the most distant star within the Milky Way. For the intergalactic distances of the *far* realm, we use either millions or billions of light-years.

The differences between the close, middling, and far realms of space can be illustrated with periods (or full stops for Irish, British, and other non-American readers). Here is our solar system and the nearest star:

.   .

The Sun and all the familiar planets, from Mercury out to Neptune, are well within the first period. If the nearest star to us has its own planetary system (we've recently discovered that it does), then all that stuff is within the second period. The nearest galaxy, thought of as an extension to this picture, would be a few kilometers / miles off the edge of the page.

Now, shrink our entire Milky Way galaxy so that it, with its billions of constituent stars, becomes a period. We can use the same mental picture to compare the distance between it and the Andromeda galaxy, as follows:

.   .

Note that this time the distance between periods is smaller, but it's still many times greater than the diameter of each. Also note that, with our galaxy collapsed, the close and middling realms are both now within the first period.

Another way of thinking about the difference between the middling and far realms of space is this. As you look up into the night sky and see individual stars of our own galaxy (like those of the arrowhead) and, close to them, a separate galaxy, consider the arrowhead stars and all the others of our own galaxy as raindrops on the windshield of a car in which you are driving along a country road toward a farmhouse light (representing Andromeda) that is just visible on the horizon.

Okay, enough of full stops and raindrops. The purpose of this exercise has been to provide a mental framework on which we can hang new names, in a sense putting them in their place and rendering them non-threatening. Thus, for example, it's enough to know that the arrowhead stars are in the realm of the middling; we can forget about their individual names and distances for most purposes. Likewise, it's enough to know that the Andromeda galaxy is in the realm of the far. That way, we can get an intuitive feel for that crucial third celestial dimension, the one we can't actually see. And we can appreciate that if we were on the surface of Mars, in the realm of the close, we would be so near to Earth that the use of our big W in the sky to find Andromeda would be almost exactly the same as from our normal vantage point on Earth.

Now let's have a closer look at what those Andromedan scientists saw when they used their amazingly advanced telescope yesterday to look at the Earth—*Homo habilis* and other early humans.

Chapter Two

Handy Man and Other Early People

From Movies to Reality

If you've seen Stanley Kubrick's cinematic masterpiece *2001: A Space Odyssey,* you'll probably recall very clearly a sequence that occurs not long after the start. An "ape-man" throws a long bone high into the air, and as it spins it transmutes into a spacecraft—thus telescoping the whole of human tool use, toolmaking, and artifact production—more than 3 million years of it—into a few dramatic seconds of movie.

The making and use of tools has been a key aspect of human evolution. We seem to have progressed from using things (e.g., bones) that were already available as tools to modifying things (e.g., stones) to make better tools (e.g., axe-heads) to the complex crafting of artifacts of all sorts, initially by hand and, much later, by machines—which are of course tools themselves.

But wait a moment. There's a risk here of falling into an ever-present trap awaiting those who try to provide a simplified account of human evolution. Our changes over time have not taken the form of an unstoppable upward linear trend in brain size, tool use, and erectness of posture leading from ape ancestor to ape-person to human being. Evolution is much messier than that. Generally, the evolution of any group, humans included, takes the form of a tree (or bush) with lineages diverging from each other for a host of reasons, including the vagaries of migration from place to place on a heterogeneous planet.

The trouble with reality, though, is that its complexity can be off-putting. An oversimplified view is unhelpful in one way, a horribly complicated view unhelpful in another. So let's try to steer a middle way between the two in relation to human evolution.

Depending on whom we believe, the number of transitional species between our last shared ancestor with chimps, about 7 million years ago, and modern *Homo sapiens* has risen from "a few" (when I was learning this stuff as a student four decades ago) to 15, 20, or even 25. But the last thing we want to do here is to bury ourselves under 20 or so species names and descriptions. No, we'll take a different approach to human evolution here, helped by our Andromedan friends.

An Alien Perspective

Because the Andromeda galaxy is about 2.5 million light-years from Earth, when the Andromedan scientists we met in Chapter 1 were looking toward us yesterday, they observed a terrestrial scene that played out some 2.5 million years ago. Let's start with what they saw and then project our view both backward and forward in time from that entry point. So we're also taking a "middle way" to approach human evolution in the literal sense of actually starting in the middle.

Around 2.5 million years ago, one lineage of protohumans was *Homo habilis*. This species is often referred to conversationally as handy-man; but here I'll use the gender-neutral equivalent, handy person. The adjective handy is used because of the abundant evidence of toolmaking by this species. At one time it was thought that perhaps handy person was the first toolmaker (as opposed to a mere tool user), but we've recently found evidence of toolmaking by protohumans more than 3 million years ago, well before the earliest handy-person fossils.

Notice that I just referred to *Homo habilis* as "one lineage," not "the lineage" of protohumans that existed about 2.5 million years ago. It's hard for us to picture a world where there is more than one species of what might be called "people." In the present-day world *Homo sapiens* is the lone representative of people; but in the world of *Homo habilis* there were at least two, and probably more. This raises the question of how we are related to them—and indeed how they were related to each other.

It seems clear that all the species of protohuman arose from a single humanizing lineage that split from the chimp lineage about 7 million years ago. So we're all related to some degree. Exactly which species begat which other species is an issue that is still taxing the best minds in paleoanthropology. Here we'll take the view that handy-person's ancestry lies in the extinct species of southern apes called *Australopithecus afarensis*. Even if this turns out not to be true, what follows regarding the evolution of brain size is affected remarkably little.

From One Brain to Another

All the very early species of protohumans, including those referred to as southern apes, had brain sizes smaller than about 500 cubic centimeters (cc). This "marker volume" is the same size as the engine of a Fiat 500. I find it helpful to think of car engine sizes when dealing with brain sizes, as they provide the most common context for the use of cc to measure volumes for the non-scientist. Handy person had crossed this threshold, though not by very much. Its brain size is thought to have been in the range 500–800 cc.

Now let's go forward rather than backward in time. More specifically, if we move to a mere half million years ago, we find a species called *Homo heidelbergensis*, named after the German city near which some of its fossils have been discovered. This species was probably also on the ancestral line to modern humans, though it's always important to stress the fact that our views on exactly who was ancestral to whom might yet change due to future fossil finds.

The brain size of this species was in the range 1,000–1,350 cc, the latter figure being also the average brain size of modern humans. Of course, we can now use liters instead of cc if we prefer. The 1.3-liter marker is interesting, as it is a common engine size in cars, but also a value within the ranges of brain size in both *H. heidelbergensis* and *H. sapiens.*

It should now be clear that although human evolution is complex and treelike, the same as the evolution of any other group of creatures, we've managed to retrieve a line, or lineage, by starting with handy person and focusing on just a few species, including one of its ancestors and two of its descendants. In a sense, we've mentally climbed our own evolutionary tree using a single route, or branch, corresponding to our special interest in the origin of *Homo sapiens.* We've ignored the other branches, but that doesn't mean they weren't there.

By the way, brain size is a very blunt instrument for measuring mental capabilities such as intelligence. It's a start, but only that. Also, we should remind ourselves that each species of protohuman had a range of brain sizes, not one specific size. And again, some individuals fall outside the range. The reason for this apparent contradiction is that the range of brain sizes quoted is usually one that applies to adults, whereas some of the fossils that have been discovered are juveniles. We'll come back to the developmental dimension of brain size soon.

Out of Africa

"Early" human evolution, a phrase I'm using for everything up to handy person, was different from its later counterpart not just in being below the 500 cc brain size marker but also in being restricted to Africa. However, after about 2 million years ago the evolution of protohumans began to be a much more global affair. Various species spread out of their African cradle, generally via the Middle East, and colonized much of Europe and Asia. However, of the species that did so, only our own *Homo sapiens* has survived thus far. Earlier African exoduses have been followed by extinctions. This applies to the well-known Neanderthals, and also to "Peking man," which belonged to different species of *Homo*.

So, geographically, Africa was our origin. This seems to have been established beyond reasonable doubt. But today we humans are found not just on the six habitable continents but also, thanks to technology, on Antarctica, whose land surface is not the permanent home of any species of mammal. We came from extremes of heat but have become able to cope also with extremes of cold.

The spread of our own species, *Homo sapiens,* was very recent when seen in the context of the whole of human evolution. Our ancestors are thought to have migrated out of Africa as recently as 100,000 years ago. When we're dealing with past eras that are measurable in thousands rather than millions of years ago, it's easier to feel the connection with the present. Smaller numbers of thousands project us forward: the first human cities were built in Mesopotamia (now Syria, Iraq, and Iran) about 10,000 years ago; almost 1,000 years ago, those Frenchmen that we call Normans, because many of their ancestors were Vikings or other Norsemen, invaded England. But we're all Africans at heart.

Feeling the Presence of Protopeople

One of the challenges I've set myself in this book is to try to collapse large distances in space or time so that we can *feel* the presence of creatures who live a long way away, or who lived a long time ago. In the case of any intelligent aliens alive today in our own Milky Way, in neighboring Andromeda, or in galaxies that are further afield, the trick is to try to mentally capture their simultaneity with us. As you read these lines, they are perhaps eating, sleeping, or walking to work. There may be a large void between us in space, but there's none at all in time.

To feel the presence of those protohumans whom the Andromedan scientists were watching last night, the trick is either the same or the opposite, depending on how we look at it. What we need to do now is to ignore the large gulf in time between ourselves and handy people, and concentrate on their closeness in space. So if we travel to the Olduvai Gorge in Tanzania, where many fossils of this species have been found (a journey of only a few hours in our time but much longer in theirs), and stand on a rocky outcrop that affords a panoramic view over the surrounding terrain, we are probably standing in the exact spot where at least one member of *Homo habilis* stood about 2.5 million years ago.

In each case, whether aliens or handy people, concentrating on the dimension—time or space—in which we are close to them is a useful technique to counteract the mind's natural tendency to concentrate on the other dimension, the one in which we are far apart. In both cases the use of this technique brings the distant creatures into full view so that we can see them in our mind's eye and picture what they may be doing—in one case right here and in the other right now.

There is, however, an important difference between imagining the aliens of today and the human ancestors of the past. The case for extraterrestrial life is based not on evidence (at least not yet) but on probabilities. If there is a one-in-a-billion chance that there is humanoid life on an unknown distant planet, and there are 100 billion such planets in a galaxy, then there should be about 100 different types of humanoid out there. In contrast, we actually have the fossilized bones and stone tools that belonged to our African ancestors.

Evidence is at the heart of the scientific endeavor, and without it we make no real progress. However, speculation is also at the heart of science, despite the disreputable (to many scientists) name of this mental activity. Indeed, speculation is logically prior to evidence. To see this, it's only necessary to consider the other, nicer names for speculation: wonder and (if it's rather specific speculation) hypotheses. Our natural curiosity leads us to wonder, and the wonder may then firm up into ideas about specific possibilities (hypotheses), which we then gather evidence to try to test. Don't let anyone persuade you that science is a simple search for evidence rather than the richer, many-sided endeavor it actually is.

What Were Our Ancestors Thinking?

We arrived at handy person by asking what those Andromedan scientists would have seen if they used their incredibly powerful telescope to look directly at us last night. Let's now engage in a little time travel (which I suspect is not possible) and sit among a group of *Homo habilis* in the Olduvai Gorge. What would we (and they) see if we had access to Andromedan telescope technology and could focus in on the Andromedan scientists' home planet? You know the answer by now: we would see their ancestors of 5 million years before the present. We have no idea what these would look like, though we can speculate that the course of their evolution might not have been so very different from ours.

As we prepare to time-travel back to the present, we might well ask ourselves this question: what would an individual handy man or handy woman have been thinking when he or she looked up at the stars? They wouldn't see even the ancestors of today's Andromedan scientists if they looked toward that galaxy, since they didn't possess any telescopes at all, let alone amazingly advanced ones. That galaxy would be a fuzzy blob to them, just as it is to us (though they'd need a variant on our arrowhead technique to find it because the shapes of constellations change over time periods of millions of years).

So, as they looked toward the stars, they would see simply stars, in the same way as we do without using telescopes or binoculars. They'd see the same stars as we do, but much more clearly than we can see them from any town or city on the present-day Earth because theirs was a world without light pollution.

The question then becomes: what did *Homo habilis* make of the stars? Did our distant ancestors, like some of our much more recent ones, think that stars were holes in a dark celestial sphere through which we could glimpse the light of Heaven? Did some of them think that the stars might be other worlds? Or did they not think in an abstract way at all? Perhaps the Moon, the planets, and the stars were all just sources of valuable nighttime light to be used for their nocturnal activities.

Although we can't answer these questions, the exercise of asking them is an interesting one. It leads to related questions having to do with the various animal cousins with which we share today's globe. Do chimps contemplate stars? There's no reason to jump to the conclusion that they don't. Do dogs? I rather doubt it but can't be sure. What about mice? Almost certainly not. And nocturnal moths? We can probably agree that astronomical issues do not trouble them much. This line of conjecture leads to acknowledging the probability that animal consciousness is a continuum rather than an all-or-nothing entity. And its evolution is still, to a large degree, a mystery. Our own individual consciousness is likewise a continuum; we'll deal with this in Chapter 3. But before we leave this chapter, there's one more thing to consider.

The Middle of Nowhere

We've already visited some strange places and times, including Andromeda in the present and the distant past here on Earth. But now let's take a vantage point that is stranger still. We'll position ourselves in intergalactic space at the point halfway between our own

Milky Way galaxy and the Andromeda galaxy, at a time that was 1.25 million years ago. The reason for visiting such an inhospitable "place-time" is to capture a fascinating encounter between two beams of light.

The beam that hits us from one direction consists of light that left our Milky Way during the era of *Homo habilis*. It sails on past us and is picked up by that powerful Andromedan telescope in another 1.25 million years, revealing our distant ancestors to the Andromedan scientists. The beam going past us in the other direction originated from Andromeda 1.25 million years ago and will be received on Earth in our present. So this is the fuzzy blob that we see in the night sky by using that convenient arrowhead composed of the three stars Caph, Schedar, and Navi. This chapter is thus ending where the first one began—that same big W in the sky.

Chapter Three

A Human with No Nerves

Changes in Our Brains

Neither your brain nor mine stays the same for very long. Indeed, the essence of brain-ness is change. I suppose you could extend that statement to equating change with the essence of life in general: there's a huge difference between the rates of change of internal parts going on in any living cell and those taking place within an inert object such as a rock. But I'd like to distinguish between two types of change in our brains over time—one of which typically occurs in seconds, the other in years.

The *seconds* timescale applies to thinking. Right now, as you're reading this, those junctions between one brain cell and another that we call synapses are firing rapidly, enabling you to take in and process information. The *years* timescale applies to how your brain got to be the size it is now (about 100 billion cells) from a starting point of zero cells when you were very young—so young that your age would be negative if we used the normal point of reference, birth.

But let's use a different reference point—conception. This is the instant that *you* began, with the fusion of a sperm cell and an egg cell. From that moment, for about the first two weeks of your life, you had no brain, and indeed no nerve cells of any kind. Evolution took animals on a journey from no brains to big brains that lasted about 700 million years. The process of embryogenesis (and post-embryonic development) has powered our own individual journeys from the same start point to the same end point in about two decades. However awesome evolution may be, surely individual development is even more so?

Let's leave this philosophical question hanging and proceed in a more productive direction. It would be wise to consider cells in general before we consider nerve cells and brain cells in particular. There's nothing like having a broad base on which to build.

The Building Blocks of Our Bodies

Given that your body, mine, and those of all other humans are made up of tiny units called cells, these should be very familiar things to us. And yet they're not, at least in terms of firsthand viewing for a person who is not a professional biologist. Yes, we know they exist. And yes, cells have made their way out of the jargon-laden world of science and into everyday speech, for example in expressions like "he doesn't have two brain cells to rub together." But the problem is that cells are for the most part too small to see. The largest human cell is *probably* the quasi-spherical egg cell, yet even that is only barely visible—it's about a tenth of a millimeter in diameter. Nerve cells can be much longer than that but they're extremely thin. Hence the "probably": it's hard to compare the sizes of cells with very different shapes.

The adult human body is made up of many trillions of cells; there are more cells in your body than there are stars in the Milky Way. The brain accounts for about 100 billion of them. And there are further millions of nerve cells in the spinal cord and in the peripheral nervous system, which connects to every extremity of our bodies.

If what I've just said is true (which it is), who are these mysterious humans with no nerve cells at all? I've already given the answer in passing: very young ones. But let's dig into this rather brief answer and dissect it a bit. We need to *feel* the presence of these incipient humans, just as we *felt* the presence of aliens (Chapter 1) and our *Homo habilis* ancestors (Chapter 2).

After a few days of development, a human embryo consists of tens of cells—the stage that's referred to as a morula, which comes from the Latin word for mulberry, a fruit that looks a bit like a raspberry because it's a multiple fruit with lots of constituent spherules. If we think of these spherules as cells, we see the connection with early-stage embryos (see the illustration on page xii).

Counting Nerveless Humans

The mulberry stage lasts for about a week, roughly from day 4 to day 11 after fertilization. During this time it increases in cell number from about 16 to about 100. At no point during this early stage of our development are any of our cells nerve cells, though of course some can be thought of, later, as having been the *ancestors* of nerve cells. So at the very least the number of humans with no nerves at any point in time—for example, *now*—is the number of human mulberries. But how many is that?

We can get a rough estimate as follows. The total human population of the world is somewhere around 7 billion. We can divide it up into progressively smaller age classes. Starting with decade-based classes (e.g., all those people between 20 and 30 years old), there are about a billion people in each. Of those, about 100 million will be within a single year-class (say, age 20–21). In each month of that year (say, March birthdays) there will be about 10 million, and within one week of that month about 2 million. Taking a similar approach to unborn humans as we just did for twentysomethings, that's the approximate number of unborn humans with no nerves who are alive today, given that the mulberry stage lasts about a week.

But there are a few provisos. (If you aren't interested in them, skip to the next paragraph.) The distribution of people between age groups in any population of humans is not even. We can represent this distribution as a diagram in which the vertical axis is age and the span of the "block" for each age category (e.g., 20s, 30s, 40s) is the number of people alive in each. Usually the diagram is pyramid-shaped, with the highest numbers present at the youngest ages and a gradual diminution to the oldest. For example, there are more twentysomethings at any one time than there are octogenarians. Therefore, the number of human mulberries is higher than our rough estimate of 2 million. In addition, the stages before and immediately after the mulberry have no nerves either, which will distort our figure in the same direction. So probably 2 million should be regarded as a *minimum* figure.

Where are all these nerveless incipient people? Where is the nearest one to you right now? In a city center, she or he is probably within a radius defined by how far you could throw a smallish stone. Of course, due to the internal nature of human embryonic development, we never see them. With other types of animals that have external development, the hiddenness of early embryos is not an issue, but small size remains an obstacle to viewing. So we have very little direct evidence of the existence of nerve-free embryos of humans or other animals, but we know they exist. And, following from the ballpark calculation above, we have some idea of their huge numbers. We also know that each of us *was* a morula (or mulberry) many years ago, for there is no other track through time from fertilization to adulthood than the one that goes via this stage.

How Does Development Work?

So here's an interesting puzzle. From a starting point of a mulberry-like thing, wherein all the cells look pretty similar, how does this tiny brainless creature begin to make nerve cells? How on earth does it know what to do? Or, since it doesn't really *know* anything (you can't know things if you don't have a brain), how does the process of embryogenesis work? How does it almost always spurn the millions of wrong ways to make a human and choose the right way? Why does our brain always end up in our head and not in our left foot?

You may be thinking that the last paragraph consists of an overindulgence of questions. But if so, you'd be wrong. In fact, the number of questions the embryo has to ask—and answer—is vastly greater than the five I've just posed. However, for the purpose of studying development we can group different embryonic decisions together into those of the same type. This makes the job of trying to understand a very complicated process a bit easier. The mulberry has to do three main types of things, as follows.

First, it has to grow, which means that it has to multiply its cell number hugely. You cannot have a brain of 100 billion cells if you only have 50 cells in total. Second, it has to make different *types* of cell, for you cannot have a brain if you have no nerve cells. Third, it has to connect these two things. The multiplication of cells and their differentiation into various cell types have to be coordinated with each other so that, among other things, the brain does end up in the head (and the toes in the feet).

Here we'll focus on just the middle one of these three things—making different types of cells—and we'll look at just a single case of it, the making of the first nerve cells. This, after all, is the beginning of the route to our brains. Strange to think that so many of us know so little about it. Let's take a small step toward solving that problem.

Back to the mulberry, then. As it grows beyond about 100 cells, it gradually ceases to be a mulberry and turns into something else. Although the continuous process of development flows like a river, we try to understand it by inventing names for different stretches of the river, all the while knowing that one merges seamlessly into the next. We use natural features of the temporal topography of development to do this, just as we might separate "Eel Stretch" from "Trout Strait" using as a boundary a right-angle bend. After a couple of bends, the mulberry ends up as something called a gastrula. This is bigger, has a lot more cells, and is a different shape—hollow, with a cavity in the middle.

The gastrula gets its name from the same source as gastronomy—the stomach. Not that it has one, but it has cells that are destined to become one. In fact, the gastrula has something the mulberry didn't: a clear division of its cells into three types. They're called the inner, middle, and outer "skins," or endoderm, mesoderm, and ectoderm. It's the first of these—the endoderm—that will go on to form the gut, which is on the inside. But the "skin" that's of most interest to us, given that our focus is on the origin of the brain, is the ectoderm (initially on the outside); all of our central nervous system comes from this source.

However, the gastrula still has no nerve cells, only their ancestors. How does the track through time of an individual ectodermal cell get influenced so that it becomes a nerve cell and not something else? The short answer to this question is *signaling*—from one cell to another—but there's a wealth of detail hiding behind that single, deceptively simple word. I'm going to evade most of that detail but still try to convey the essence of the process; this is a considerable challenge, but it's one worth rising to.

On Becoming a Small Sausage

When you were a gastrula you were sausage-shaped. The long dimension of your tiny sausage was your head-to-tail axis. It will help here if we consider a sausage that has been blackened by overcooking on one side but not on the other three (not that sausages have a square cross section, but you'll see what I mean). Now position the sausage on a plate, burnt-side up. Inside the sausage and running parallel to its long axis, imagine a thin rod of denser tissue, perhaps something like cartilage, which lies immediately underneath the burnt (upper) surface. It wouldn't be nice to eat, but this sausage is for explaining things, not for eating.

In an embryonic sausage the burnt surface is caused not by overcooking but rather by signals from the thin rod underneath. These signals say "become dark brown cells." And in fact "dark brown" in this context means "nerve." That's how it works. Later on the nature of the signals changes to instruct the nerve cells to do various things, including to become different types of nerve cells—for example, some become motor neurons, those that go wrong in the affliction called motor neuron disease (also called ALS), of which the most famous sufferer is the great physicist and cosmologist Stephen Hawking.

The signals are many and varied. However, an important player is sonic hedgehog. But I haven't strayed into the realm of fiction here. Sonic hedgehog is very different from its near namesake, the video game character Sonic *the* Hedgehog—note the sandwiched definite article, which makes all the difference.

Sonic hedgehog is a protein, albeit one whose name was arrived at with a sense of humor. We ingest proteins on a daily basis, break them up in our guts into their component parts, and then reassemble these parts into different proteins in the rest of our bodies. One of these latter proteins is sonic hedgehog. However, when embryonic proteins are made, we—the little sausages—are not getting food from eating, we're getting it via our mother's blood, ever since the time that the initially wandering embryo implanted itself in the uterine wall. But that's just a detail—it's our mother who is eating, not us.

In our bodies—whether embryonic or adult—most proteins take the form of tightly tangled balls of string. We can picture molecules of sonic hedgehog as balls of bright blue string to distinguish them from other proteins, which all have different colors and hues. The thin rod within our embryonic sausage secretes these blue balls. In the tissue immediately overlying the thin rod (picture this rod as being blue too, just to mentally distinguish it from the rest of the sausage) there is a high concentration of blue balls. Further away, the concentration subsides. Ectodermal cells receiving sonic hedgehog balls react not just to receiving them but to how many they receive, and this, among other things, determines what type of nerve cell they become.

You can picture this process going on indefinitely as a sort of relay race. Once a new type of cell has been formed as a consequence of "upstream" signaling, it in turn can go on to do some signaling of its own, and hence cause further "downstream" processes to occur in the embryo. Not all cell types do this, but some of them do, and that's enough to take development all the way through to its completion. It's complex, but in the majority of cases it works. And when mistakes are made, they're usually minor. Polydactyly (extra fingers), rare as it may be, is commoner than "polybrainy," which is either vanishingly rare or fictional, depending on your point of view.

So as we go forward through embryonic time, signaling between cells can be seen as an important process that drives development. But going backward, we encounter a problem. At the mulberry stage, all the cells seem more or less the same. So how can any one of them start making a signal while the others just end up receiving it? In other words, how does the process of cell differentiation *begin,* especially when all the cells contain exactly the same genes? It looks like a hopeless case.

But it's not. In fact, there are many ways in which an initially homogeneous embryo can begin to regionalize and make different cell types. One was proposed by the famous computer scientist Alan Turing, who also made important contributions to biology. Turing proposed a model wherein a ring of cells that were initially all the same could generate differences so that some of the cells would act as the starting point for tentacles, others not. Thus he could explain the origin of the tentacles that surround the mouth of the tiny freshwater creature that we call a hydra. And from the perspective of how developmental systems work, that's essentially the same problem as how we come to be sausage-shaped despite starting as a mulberry. The real reasons for the appearance of a hydra's tentacles or the shape-shifting that takes us from mulberry to sausage are more complex than Turing's model, but it was an important step in the right direction.

Beyond the Sausage

The dark strip down the dorsal side of our embryonic sausage that is our incipient central nervous system is a beginning rather than an end. This strip is initially flat or slightly curved, like the sausage-skin I'm using to portray it. It later turns up at the edges so that it becomes U-shaped. The ends of the U then grow together and fuse, forming a tube—the neural tube, as it's called in the embryo. Then, due to receipt of certain signals, the head end of the neural tube outgrows the middle section and the tail end, thus beginning the distinction between the brain and the spinal cord. And the brain cells just keep on multiplying until there are billions of them, so we can use them to write or read, as we're doing now, or for many other things. Mulberry becomes sausage becomes sentient life-form. An everyday miracle.

Now you can see that statements like "handy man had a brain size of about 600 cc" are very biased—in a way that the Italian biologist Alessandro Minelli calls "adultocentric," for example in his book *Perspectives in Animal Phylogeny and Evolution*. The brain of any human being is a trajectory through time. It starts at 0 cc and ends up much larger—exactly how large varies from one individual to another. This is as true of handy person or any of our other ancestors as it is of present-day humans. It's true of other animals too. We are all four-dimensional creatures; we have trajectories through time as well as space.

To us as observers, cells are very small things. But to a molecule of sonic hedgehog they're very big. So big, in fact, that each of these tiny blue balls in the cell is like the proverbial drop in the ocean. And yet to a chemist a sonic hedgehog molecule is *huge*. Compared to a molecule of carbon dioxide, many trillions of which you have exhaled in the last few seconds, it's an absolute monster. If a small Irish village with just a few houses around a crossroads is carbon dioxide, sonic hedgehog is New York.

Our bodies are made up of countless molecules, some large, some middle-sized, and some small. Each of these in turn is made of atoms that belong to an assortment of elements—for example, carbon, oxygen, and iron. Biologists trying to explain the near miracle of embryogenesis don't usually spend too much time thinking about where the elements come from. But astronomers do, because in the early universe none of them existed. So, where *did* they come from? Let's find out.

II

CYCLES OF LIFE

Chapter Four

From Celestial Furnaces

Going Down

It's not hard for you to take a photo of a friend. We take it for granted that people are visible—and hence photographable. But what if we want to photograph only part of a human? How hard that is depends on how small the part is. Portrait photos are commonplace, so the head's not a problem. But suppose we want a shot of an individual organ? The eye is an example. Try *human eye* in Google Images and you'll see some fine photos. Many of our other organs—heart, liver, brain—are larger and hence also potentially photographable. But the fact that they're internal organs poses a difficulty, so we only see photos or footage of these in dire circumstances—(fake) hearts in mid-operation on a TV show, or (real) brains in jars of preservative in museums, when the owners are no longer with us.

Then, in "going down" to progressively smaller parts, we reach an impasse. Organs are made of cells, but cells aren't big enough for their likenesses to be captured with an unaided camera. Instead, anyone wanting to obtain meaningful images has to mount their camera (and not an everyday compact, like mine) on top of a microscope. They are then practicing the art-science of photomicroscopy.

This can produce stunning images. Most of us can mentally picture various human cell types, such as nerve cells, sperm cells, and blood cells, because we've at some stage seen professional photographs of these cells, or diagrams that are based on them.

Going down further, we reach another impasse. As we saw in Chapter 3, cells are made of molecules, and these in turn are composed of atoms. In this realm the ordinary microscope becomes as useless as the ordinary camera already did one stage earlier. However, we can obtain images of big molecules such as DNA with specialist techniques; and recently, small molecules and even individual atoms have been imaged. In 2013, an international team of scientists obtained the first photograph of a hydrogen atom, the smallest of all, using a technique called quantum microscopy, which sounds fictional but is in fact a real scientific method used on present-day Earth (and perhaps elsewhere).

At this point we should pause and ask: why the obsession with photography? Why do we need to see things in order to understand them? I'll let someone else give you the answer: one of my foremost scientific heroes, Thomas Henry Huxley. He was the man who became known as "Darwin's Bulldog" for his fierce defense of the theory of evolution against various adversaries, most famously Bishop Samuel "Soapy Sam" Wilberforce. In *The Crayfish,* published in 1881, Huxley writes that "those who read without acquiring distinct images of the things about which they read, by the help of their own senses, gather no real knowledge, but conceive mere phantoms and idola" (5).

In the case of atoms, the theory came before the images. But the images have helped to confirm that atoms are no less real than organs. They're just a lot smaller. We humans, our fellow animals on planet Earth, the Earth itself, the Sun, and, as far as we know, all other suns and planets in the universe are made of atoms (albeit these exist as ions when at high temperatures within suns). So too are Andromedan scientists made of atoms, together with all life-forms everywhere, of which there are probably countless types.

Although this is true right across the observable universe today, it wasn't always so. In the beginning—say, the first few seconds after the big bang, through which the universe was born about 14 billion years ago—there were no atoms or molecules at all. So how did these tiny things that are the ultimate building blocks of humans and all else originate?

A Productive Half Hour

There are two answers to this question, one involving the early universe, the other (treated in the next section) involving stars. The first of these stretches our credulity to the limit, and Huxley's warning about "phantoms and idola" should be kept firmly in mind. Nevertheless, what follows is currently believed to be true.

In the first second of its existence, the universe contained no atoms, nor even most of the atomic nuclei (consisting of protons and neutrons) on which atoms are based. These formed later, over a period that lasted about half an hour. When the universe was 30 minutes old, atomic nuclei were everywhere. The process that produced them is called primordial nucleosynthesis, to distinguish it from the later process of stellar nucleosynthesis that occurs in stars.

At the end of this phase of primordial nucleosynthesis, there were atomic nuclei in the trillions (*gazillions* would probably be a better label, even though it's undefined). However, despite their almost infinite number, these nuclei were very limited in type. Nearly all of them were either hydrogen or helium—the two elements of which the Sun and other stars are mostly composed today. Our bodies contain a lot of hydrogen. So we can see, in one sense anyhow, where that came from. But our bodies also contain a lot of carbon atoms, together with many others such as nitrogen, oxygen, and iron. To see where these came from, we'll have to go inside a sun.

Before we do, though, there's one final thing to take care of here. You may have noticed that I casually slid between atoms and their nuclei in the above account. But the slide of reality took longer than my slide of description. Atoms themselves—that is, the nuclei plus their outer shells of electrons—didn't appear in their wholeness until nearly half a million years later than the electronless nuclei. We'll consider that issue further in Chapter 7.

Inside the Sun

At the start of the book I suggested that you step outside on a clear night and gaze at the stars. Now I'd like you to imagine stepping outside on a clear day and looking at the Sun. (Of course you should never look directly at the Sun, as it can damage your eyes.) This amazing object dominates our daytime sky when there aren't any obscuring clouds—rather often in California, and rather rarely in the Emerald Isle.

Actually, in one respect it doesn't much matter whether you consider the Sun or a star because they're the same sort of thing: the stars are distant suns, functioning in much the same way as our own. But somehow it's easier to picture our impressively bright Sun as a gigantic nuclear furnace than it is to picture what appears to be a delicate little distant star in the same way.

In the heart of the Sun, hydrogen is being converted into helium by the process of nuclear fusion. This works through Einstein's famous equation $E=mc^2$. That is, the amount of energy (E) that you get by destroying a piece of matter in the heart of the Sun is its mass (m) times the speed of light (c) squared. Since the speed of light is very great, you get an awful lot of energy from destroying a small amount of matter.

What happens inside the Sun, then, is that hydrogen nuclei are smashed together at incredibly high temperatures (many millions of degrees on whichever scale you use). The result of this is the formation of helium nuclei. If you look at the array of things that go into this reaction and the array of things that come out, you find that the latter are, together, a little lighter than the former. The difference has been converted into energy—lots of it. According to NASA, the Sun produces enough energy to completely melt a pillar of ice that's thicker than a skyscraper and as long as the distance from the Earth to the Sun—in a single second.

Dying Stars

So, lots of hydrogen was created in the early universe. Quite a lot of helium was produced then too, and continues to be produced inside stars like our Sun. But that doesn't help us much to understand where elements such as carbon, of which our bodies are made, came from. These are not being produced by the Sun. To discover how these crucial components of our bodies are made, we have to examine the internal chemistry of stars that are no longer living, as the Sun is, but dying.

A digression: Chemistry was one of my worst subjects at school. Biology and physics I loved, chemistry I hated. The sight of a wall poster of the periodic table of the elements had me dashing for cover. Boredom at first sight, I thought then. But boredom is a subjective thing. One person's eureka is another person's yawn. And the two people involved in the comparison can be different in space or time. In this case it's time—and the two people are me as a schoolboy and me in the present day. In case you in the present day are like me in the past with regard to your perception of chemistry, what follows is intended to be fun.

Atomic nuclei are made up of protons and neutrons—picture these as ping-pong balls, with the protons (positively charged) being red and the neutrons (no charge) green. The simplest of all nuclei is that of hydrogen—one red ball and no green ones. Helium has two of each. Carbon, that all-important element for life on Earth, and probably elsewhere, has six of each. And getting toward the upper end of the size spectrum of elements, uranium has 92 red balls (protons) and about 146 green balls (neutrons). The reason for the "about" is that all forms of uranium are unstable and undergo radioactive decay into something else. You can get radioactive carbon too, but the form of carbon I've mentioned above (with six plus six balls) is the main non-radioactive one.

Each element is uniquely determined by its number of red balls. Of the elements mentioned so far, hydrogen has 1, helium has 2, carbon 6, nitrogen 7, oxygen 8, iron 26, and uranium 92. The highest number known in the mid-1990s, 112, belongs to the unfamiliar element copernicium, named after the Polish astronomer Nicolaus Copernicus, the man who proposed a Sun-centered solar system in the year 1543. Today, the highest number of red balls known has risen further to 118 (oganesson, named after the Russian physicist Yuri Oganessian).

When I saw the full range of elements in the form of a large tabular wall poster with a lot of words and numbers on it, I think the boredom-inducing thing was the feeling of being drowned in detail. This is why I've mentioned only nine elements here, instead of all 118 of them. It's much more interesting to look at a small sample rather than the lot, and then shift our focus to how they originated, which is where we're going in a moment after a few thoughts on the physical sizes of atoms and ping-pong balls.

If you had a carbon nucleus made of 12 ping-pong balls (6 red and 6 green) sitting on a table in front of you, it would be bigger than your fist but smaller than your head. That's a good size of model to use if you want to picture it clearly in your mind. But a model is just a means to an end. What size is the real thing—the carbon nucleus—that your red-and-green ball of balls is helping you to envisage?

To give a more satisfying answer to this question than the obvious "very small," we'll proceed as follows. Sitting beside the ball of balls on the otherwise spotlessly clean table is a grain of sand. Like most grains of sand, it's very small—probably about a millimeter across. And in terms of shape we'll imagine it to be quasi-spherical, but a bit rough around the edges, for sure—this is not the perfect sphere that you find in the world of geometry.

At this stage in our thought experiment, the ball of balls is a lot bigger than the grain of sand. But now let's imagine shrinking it, and not just a little but a lot. Suppose its initial diameter is 10 centimeters (or 4 inches), even though we would have to use a nonstandard ping-pong ball to achieve this precise girth. So before shrinkage the ball of balls has a diameter 100 times that of the grain of sand that's sitting close to it on the table. If we shrink it to 1/100 of its original size, it's the same girth as the sand grain, so you can still see it. You might still just catch a fleck of red or green.

Now do the same again—reduce the size by another factor of 100. It's invisible, but it's still there. Now repeat the shrinkage again. At this stage, the ball of balls is a millionth of its original size. Is that small enough to be about the size of the carbon nucleus that it started life as a visible model of? No, we'd have to repeat the 100-fold shrinkage several more times for that to be the case. Welcome, briefly, to the world of the atom. Now, back to the astronomical world, the world of the very big—but how big?

Age Matters

The familiar expression "size matters" applies to stars. So does the (less familiar) "age matters." And indeed for stars these two things are related in an interesting way. Big stars burn hotter and live shorter lives. They also die in a different way than small stars do. For our purpose here—understanding where carbon and other elements in our bodies come from—star death is much more interesting than star life. For most of its life, a star simply makes helium from hydrogen, and that's true regardless of its size. Since there is no, or nearly no, helium in our bodies, this doesn't help us much. The heavier elements—in other words, the ones that have more red and green balls in their atomic nuclei than helium has—are mostly made in very old stars that are in the process of dying.

But what is "old" for a star? Our Sun is a little under 5 billion years old. We believe it to be a middle-aged star, about halfway through its life. Ten billion years is a long life, and the Sun is privileged to live this long on account of its relatively small size. Compare this with a star that has 25 times the mass of the Sun. The large star will live for only about a thousandth of the time the Sun does: about 10 million years rather than 10 billion. In fact, the life span of this large star may be closer to 7 million years than 10 million. That's the same as the time it took humans and chimpanzees to evolve from our last common ancestor—a tiny fraction of evolutionary time here on Earth.

At first this contrast between big stars and small ones seems counterintuitive. Shouldn't it be the other way around? After all, big stars begin life with more nuclear fuel, so surely it should take them longer to exhaust it? Well, if other things were equal, this would be true. But they're not. The cores of big stars reach considerably higher temperatures than those of small stars. This means that their nuclear furnaces burn faster, and this more than outweighs the difference in the initial amount of fuel.

Regardless of life span, all stars eventually die. Our Sun will follow the normal trajectory of a smallish middle-aged star and will eventually swell to become a red giant and then collapse to form a white dwarf. In the process of this collapse, it will shed a great cloud of gas and dust into space. (This cloud is an essential ingredient for life, as we'll shortly see.) A star that's 10 times or more the mass of our Sun will die in a completely different way: it will swell to become a supergiant, which will eventually self-destruct in an amazingly bright explosion—a supernova—blasting material that it has made in its nuclear furnace into space, and sending a gigantic shock wave out in all directions.

In both cases the important contribution of the dying star to future life-forms is the great cloud of ejecta. The gas and dust from a supernova explosion and that originating from the relatively quiet death of a star the size of our Sun contain many heavy elements, including the crucial carbon that is at the heart of all living things on Earth and perhaps elsewhere too.

But we're not just carbon. Carbon alone is quite boring. It can take different forms, such as the "lead" in a pencil (a form of carbon called graphite, not lead at all) or the diamond in an engagement ring (a crystalline form of carbon very different from graphite, but still carbon). Now, admittedly diamonds are interesting rather than boring, especially if you're interested in wealth. But from another point of view, as a basis for life, pure carbon in any of its forms is really rather dull. It becomes much more interesting to a biologist—as opposed to a jeweler—when it combines with other elements to produce, ultimately, the stuff of life, including DNA.

But wait a moment. If stars just turn hydrogen into helium in their nuclear furnaces, why is the gas and dust that they ultimately eject in one way or another not just composed of vast amounts of helium? The answer to this question lies in the altered nuclear fusion reactions that go on in old stars. As a star ages, it produces a greater variety of elements in its nuclear furnace. Basically, as it gets older its core gets hotter, and the hotter it becomes the heavier the elements that can be made. In a dying Sun-sized star the temperature gets high enough for carbon and oxygen to be made, but for most of its dying nothing heavier is produced. The exceptions are dramatic but short-lived death throes that produce a variety of *very* heavy elements such as krypton. This is a gas, of which there are small quantities in the Earth's atmosphere, whose atoms have 36 red balls (protons); it is not to be confused with kryptonite, a mythical material from Superman's home planet.

In a dying star that's an order of magnitude more massive than our Sun, the temperature gets even higher. So, as well as making carbon and oxygen, such stars make many other elements, including iron. Although the name of this substance tends to conjure up mental images of iron girders, more subtle sprinklings of iron are essential to life. Right now, as at every other moment in our post-mulberry lives, blood is circulating through all parts of our bodies. It's red, as we know, even in the case of kings and queens. Each red blood cell is made up mostly of a protein called hemoglobin. A tiny fraction of a hemoglobin molecule—less than 1/1,000—is iron. If we could magic away the iron from our hemoglobin—an unwise experiment if we could do it for real rather than just in our minds—we would be dead in seconds. There's not a lot of iron in our blood, but it's the difference between life and death, as it's the iron that allows us to transport oxygen around our bodies. Without dying megastars we wouldn't exist.

But so far, this argument has a serious logical flaw in term of its relevance to human life. When our Sun dies, it will indeed shed "stuff" in the form of a star-death nebula into space, and among this stuff will be lots of carbon and oxygen. But these won't be much use to us because we'll have been dead for about 5 billion years. And if a megastar goes supernova at about the same time as our Sun shrinks to a white dwarf, it will blast some iron out into the void as it explodes; but again this will be rather too late to be of use to our red blood cells.

What was important to *us,* as opposed to carbon-based life forms of the future in other solar systems, was the death of earlier stars long before our own star and its accompanying planets were born. We are indeed made of stardust, but the stars concerned are long gone. We are here because they are not.

On Generalizing

Science is a wonderful thing. It attempts to explain fiendishly complicated processes in as simple and elegant a way as possible. And part of this endeavor is the search for general principles so that, whenever we can, we explain lots of things as subtle variants of just one essential thing, all of the variants being at heart the same and just a little different in detail. If the "things" are stars that are approximately the same size as our Sun, then we see them as having approximately similar life spans and deaths. This is true no matter where a star is, and indeed no matter when it is, was, or will be. We expect the stars of the past and the future to live, age, and die in ways that aren't too different from those of the present. In general, we expect the laws of science to apply across the whole of space and time.

Is this a noble hope and a fair expectation, or is it just human arrogance? I suspect it's the former. But we must be ever on our guard to detect signs that our proposed generalizations are wrong. I turn to my scientific hero again at this point. Thomas Henry Huxley once referred to "the great tragedy of Science—the slaying of a beautiful hypothesis by an ugly fact" (1896, 244). What he meant was that sometimes, as science progresses, we discover that an elegant general theory, which at first seemed to account for a wide range of phenomena, is wrong. Although Huxley described it as a tragedy—the death of something beautiful—he still regarded it as an essential thing. That is, although we search for elegant general theories, like Darwin's theory of evolution, we must discard them if some piece of evidence comes to light that shows they are incorrect. Scientists search for both truth and beauty, but if there's a conflict, truth must prevail.

At present we have no reason to believe that the life cycles of ancient stars should have been fundamentally different from the life cycles of current ones. Nor do we have any reason to believe that the life cycles of stars of the distant future should depart in any major way from those of the past or present. But perhaps thinking about the *life cycles* of the stars of any era in the history of the universe is unwarranted. After all, stars are not alive. Animals have life cycles that connect them with their parents and their children. But surely stars have no real interconnected lineage through time, no ancestors and no descendants? Well, in one way they do and in one way they don't. Let's now poke at this yes-and-no answer to the question of whether stars are like animals.

Chapter Five

Life Cycles: Animals versus Stars

Humans in Time

What's the longest life span you could hope for? If you're a real optimist, I'd say about 125 years (subtract five if you're male), though most of our individual tracks through time are much shorter. Right now, as I'm writing this (but perhaps not when you're reading it), the record to beat is that of a Frenchwoman, Jeanne Calment, who died in 1997 at the grand old age of 123. If you look this lady up in a list of the world's oldest people, you'll find that her age is given as 122 and (approximately) a half. My 123 comes from adding 9 months, because birth can be regarded as merely a change in environment—albeit a rather drastic one. A human life begins at conception; this is the real beginning of each individual track through time. We have the same genes from conception to death, and while genes don't on their own make a person, they do contribute quite a lot.

My advocacy of conception rather than birth as the ultimate beginning of each human life should not be taken as support for one point of view over another in the debate about abortion. That's a complex debate, and one in relation to which I claim no special expertise. I'm thinking not in a political way here, but in a scientific one. It's very clear to me, as a biologist, that the moment of conception (which is the subject of Chapter 6) is the beginning of a human life, albeit the first couple of weeks or so are spent as an early-stage embryo that has no more consciousness than an oak tree. That's guaranteed by the fact that neither the human mulberry nor an oak tree has any nerves. The only way around this apparently inescapable conclusion is if there can be consciousness without a nervous system. So far, there is no evidence for this whatsoever.

Anyhow, let's consider some human life spans that are more typical than 120+ years and think about how these connect up as parental and offspring life spans in the form of life cycles. One good approach to this subject is to ask the question: did you know your great-grandparents? Regrettably, I didn't know any of mine.

Let's pick some possible life spans and see how they affect the likelihood that we knew our great-grandparents. We'll start with an optimistic view—though not quite so optimistic as the 125 with which I began this chapter. Imagine a family in which everyone lives to be exactly 100, each woman has twins (a girl and a boy) at 25, and the earliest age at which long-term memory is effective is four. The twins of any particular generation in this family will, when they grow up, remember both their grandparents and their great-grandparents. The latter were age 75 when our twins were born; they die when the twins reach 25.

If we become more pessimistic about life spans, the situation changes as follows. If family members all live to be 75 rather than 100, our twins remember their grandparents but not their great-grandparents, who will have died at about the time the twins were born. If it's a cruel world and all family members die at 50, then the twins only remember their parents. Grandparents are known of only through photos and stories.

Looking back through time to previous generations is only half of the picture. We can look forward as well. If you knew your grandparents, they knew you. It doesn't necessarily work the other way round, though: my father knew my daughter but, because he died when she was only three, she has no memory of him, which is a pity. In the real world, of course, all families are different from each other and all members of a particular family are different from each other too. There are some unlucky people who cannot remember their parents, and there are some who can bring back distant memories of a great-great-grandmother.

So, what's the point of this exercise of looking backward and forward through time? It's intended to be a very visual beginning to an exercise of looking much further in time, in both directions, when we don't have memories or even photographs to help us. We can no longer picture the individual people concerned, but we can at least count how many generations of our ancestors it takes to get back to any milestone, time-stone, or tombstone in the past. Equally, we can think our way, generation upon generation, into the distant future, although perhaps this is less meaningful because, as they say, the future is not set.

How Long Ago Was the Past?

The period between birth and giving birth is called the generation time. In the above musings about grandparents, it was taken to be 25 years. If we retain this vaguely realistic figure, how many generations back was 1066, the year of the Norman invasion of England? The answer: 38, which to me seems a surprisingly small figure. How many generations back to the first exodus of *Homo sapiens* out of Africa? This time it's about 4,000. What about going all the way back to a particular generation of a *Homo habilis* family that lived 2.5 million years ago? Well, in that case the number of generations is somewhere around 100,000. But this is still only a third of the way back to the first protohumans and a tiny fraction of the way back to the first-ever animal—less than half a percent of that vast span of time.

Although the overall evolutionary process is a tree rather than a line, if we take any particular present-day species, whether our own or another, and trace it backward, we *do* end up with a line—a lineage, as it's called—that connects us with any point in the history of life on our planet. As we've seen, life spans connected by reproduction form life cycles. Life cycles connect us not just to our distant protohuman ancestors but to their ancestors too. A long time ago, our ancestors were fish, a fact captured by the title of the book *Your Inner Fish* by American paleontologist Neil Shubin. If you could trace your family tree back far enough, that's where it would get to (though you could go further back still). Genealogy becomes anthropology becomes evolutionary biology. Countless life cycles linking back through the proverbial mists of time to a particular pair of fish that swam in ancient oceans, with no breaks whatsoever: if there was a gap of a single life cycle, we would not exist.

So we can see how life cycles link to evolution. Let's now see how they link to inheritance, and how evolution and inheritance link to each other. Although the inheritance of traits between generations is a prerequisite for Darwinian evolution, there's an interesting difference in emphasis between short-term studies of how traits are inherited and long-term studies of how they evolve. Take height, for example. Tall parents are more likely than short parents to have tall children. The emphasis is on the similarity between generations. But we humans of today are taller than our protohuman ancestors—now the emphasis is on difference rather than similarity.

How does a process of inheritance causing similarity between generations produce, in the long term, a process of evolution causing not similarity but *difference?* There's a two-word answer to this question, and we owe it to Charles Darwin: natural selection.

You can view evolution as a sort of competition between genes that produce different values of traits such as height. At any moment in time in a human population, there are some families in which "tall genes" predominate, some in which "short genes" predominate, and others that are a fairly even mixture. If the character *height* is positively correlated with reproductive fitness, then in the long term families with tall genes will leave more offspring and the human lineage overall will get very slowly taller.

Of course, it's far more complex than the story I've just told. Your height is caused by a mixture of things: individual genes, interactions between genes, environmental factors such as childhood nutrition, and gene-environment interactions. And we can ask: what happens when a man from a tall family and a woman from a short family, or the other way round, produce children? This is the case in my own family, with the unfortunate result (for me) that I am short and my younger brother is tall. But such aspects of sibling rivalry are blips in the long march of evolution—a march of averages against which the variations of individuality are powerless.

Let's summarize the argument so far in terms of our own species. We evolved from another species of *Homo* in Africa and spread from there to our current cosmopolitan distribution. All the while, offspring have a tendency to resemble their parents, and yet in the long run we are very different from our ancestors. If this is the way a system of life cycles, inheritance, and natural selection works, it should be the same for any other species of animal on the planet. And so it is. To see this, we'll take an animal that's very distant from the human one in the great tree of life: a pond snail.

Underneath a Water Lily

I'm writing this in my study at home. In the garden not far from where I'm sitting is a pond with a luxurious growth of water lilies that have pink flowers. If I push aside one of these beautiful flowers, take hold of a leaf, and turn it over, I find several inch-long gelatinous masses adhering to the leaf's undersurface. If I look at these semitransparent globs carefully, perhaps using a magnifying hand lens, I see lots of little specks. If I take a tiny chunk of the gelatinous material containing just a single speck and put it under the very ordinary low-power microscope that lives on a shelf in my study, I see that the speck is really a miniature snail. This is close to the start of the nascent snail's life cycle; the adult snail that laid the egg from which the baby snail hatched was probably close to the end of its life cycle, given that individuals of this species—called *Lymnaea stagnalis*—often die shortly after they lay eggs.

There are both similarities and differences between the life cycle of this species and that of our own, but the similarities are more important. In both cases, a "child" receives a set of genes from both parents. In both cases, trait values in the parents tend to be reflected in roughly similar trait values in the offspring, albeit with much individual variation. In both cases, trait values shift slowly in the long term. In these snails, the shape of the shell has been much studied, and we can think about its inheritance and evolution in much the same way as we thought about the inheritance and evolution of human height. In the short term, parents with squatter shells tend to have squatter offspring, though we should note that there's quite a large environmental effect on shell shape as well as a genetic one. In the long term, the group of species to which this one belongs clearly evolves in terms of squatness or elongateness of the shell, because there are other species of *Lymnaea* alive today that differ in shape from *L. stagnalis*.

These similarities mean that natural selection works in the same broad way in snails as it does in people. But what about differences between our life cycle and theirs? Well, if you were a baby snail, both your parents would have been hermaphrodites and your birth would have been from your "mother," who might, after accepting sperm from your "father," have adopted the male role and engaged in another mating with your "father," who is now adopting the female role. Because of this, your sister's mother is the same individual as your father. Even these days, when there are plenty of transgender humans, this would be a hard act for members of our species to follow. But this difference is only window dressing on the serious business of mating, producing offspring, and completing our otherwise similar life cycles.

From Snails to Stars

Does an animal life cycle—whether of a snail or a human—have a beginning? This question is often posed rhetorically by asking which came first, the chicken or the egg. The egg-chick-hen / rooster cycle has repeated itself countless times all over the Earth and will doubtless continue to do so far into the future. It has no discernible beginning. If you trace it back far enough in evolutionary time, "hen" becomes "fish" and "chick" becomes "fishlet," because hens, like us, had fishy ancestors. For some reason that's not clear to me, I tend to think of the egg as the start of the cycle, but I'd be hard put to offer a logical justification for that point of view.

For a star, what's the equivalent of this cycle of egg-chick-hen? Indeed, is there one at all? Stars aren't animals, so why should they have life cycles? Well, animals have no monopoly on life cycles—plants have them too, as do microbes. But perhaps the more general category of *life-forms* has such a monopoly? Apparently not, for it's now common to speak of the life cycles of products in economics and business studies. This raises a crucial question: what *is* a life cycle? If we can't define it, there's a risk that it will come to refer to so many things that it will be effectively meaningless. Let's keep this issue at the back of our minds; we'll return to it shortly.

One way of looking at stars is as time sequences that run as follows: gas cloud, protostar, star. Although this is a well-established series of transitions, it's less familiar to most people than the egg-chick-hen sequence, so let's dissect it a bit. This may require us to change the way in which we view outer space.

We will now get into a hypothetical (for the moment) spacecraft that can travel at close to that universal speed limit, the speed of light. We blast off from Earth later today, heading in the direction of the nearest star to our solar system—Alpha Centauri. This is actually a triple-star system, but that's not important for our purposes here. It can't be seen from the Northern Hemisphere, but it's visible in the southern sky. It's about 4.5 light-years away, so our journey will take approximately five years—*exactly* how long depends, of course, on just how close to light speed we can get.

In reality, the fastest speed attained by a human-built spacecraft to date is around 65,000 kilometers per hour. This is considerably less than 1 percent of the speed of light. Right now, the only spacecraft that can travel at "warp speed" are the *USS Enterprise* and other inventions of science fiction. But in the future, who knows what is possible; we may get there one day.

Anyhow, suppose that we break down exactly halfway to the Alpha Centauri system. We are temporarily stuck in interstellar space, with both our own Sun and Alpha Centauri more than two light-years distant. We draw lots for who goes outside with a sonic screwdriver to make a magical fix of the problem—whatever it may be. You're the lucky winner. You make the repair but then decide that since you're out there you may as well take a sample of space to see what's in it. Conveniently, you have a small box in your pocket measuring 10 centimeters (4 inches) across. You fill it with space, pocket it again, and clamber back inside. You've sealed the box very carefully; after all, you're not going to be back home for another seven years or so.

Eventually, on return, you analyze the contents of the box. You find that, contrary to the view that space is completely empty, it has quite a few things in it—many photons of radiation, some protons (hydrogen nuclei), and a few atoms. Of course, compared with a boxful of our atmosphere, this is very rarefied indeed. Nevertheless, it's not nothing.

We now believe that space is typically like this. Rather than being an objectless void, it contains an extremely low density of small particles. And, scattered about here and there are denser clouds of particles. One of these can be seen in the constellation of Orion the Hunter, about halfway down his sword. With the naked eye it just looks like a star—though maybe a slightly fuzzy one. Through a telescope you can see a great gray-white cloud, partly obscuring stars that lie within it and behind it. Those within it—including the cluster of stars called the Trapezium—are young stars, recently born. This combination of dense gas clouds and young stars is no accident. Stars form from such clouds, or parts of them, that collapse under their own gravity.

Gravitational collapse is a sort of runaway process—once it starts, there's no stopping it. When a cloud in space begins to collapse, it gets progressively denser and hotter. When the core temperature reaches a certain threshold (a few million degrees), nuclear fusion begins and the protostar becomes a star. At this point its surface temperature, which is considerably lower than the internal temperature, is typically 2,000–3,000 degrees—about half that of our Sun. (All these temperatures are expressed on the Kelvin scale that begins at absolute zero. It is named after Lord Kelvin, alias William Thompson, a physicist born in Belfast—the same city I was born in about 150 years later.)

So, we have a rough idea of how stars are born. And we already had a glimpse, in Chapter 4, of how they die—in different ways depending on their size. Let's put these things together and sandwich between them the "main sequence" phase in which stars spend most of their lives; our Sun is in that phase at present.

A smallish star, say the size of our Sun, spends a few million years being born, then a period of *billions* of years (about 10 billion) as a stable star in the main sequence, fusing hydrogen into helium and producing copious amounts of light and heat. Then it spends a few million years undergoing the major changes that turn out to be its death throes—swelling to red giant, then casting off a great cloud of gas, with what remains in the middle shrinking to become a white dwarf.

A bigger star, say ten times the mass of the Sun, has a broadly similar beginning and type of life (though a much shorter life, as we saw earlier), but it dies in a supernova explosion that blasts much of its material into space, with the remaining material shrinking, under the effect of gravity, into a very strange object called a neutron star. In such a body, the atomic structure has collapsed so that effectively we just have neutrons (hence the name). A standard-sized sugar cube of this material would weigh billions of tons—hard for us to imagine.

From Sequences to Cycles

Now here's the crucial point. Whether we're talking about big stars or small stars, all the above information is about *sequences,* not *cycles,* in time. Sequences only become cycles if we can join their ends to their beginnings. Can we do this for stars? The answer to this question is a resounding yes. The dying of both small and large stars throws vast quantities of gas into space. This gas joins other gas already out there. And it's from this combined gas that the next generation of stars will form. Not only that, but one event that can initiate the collapse of a region of gas is the shock wave from a supernova explosion. This can increase the density of some gas clouds that it blasts through to a value that's higher than their threshold for collapsing. So supernovae not only contribute material for new stars, they can also serve as the trigger for their formation.

So, after all, stars do have life *cycles*. But, that said, these are very different cycles from those of animals. In particular, stars can have many parents, in most cases probably *very* many—too high and variable a number to count. This contrasts with animals, where the number is small and usually two. It can also be one, for example in the case of some male insects, and in rare cases it can even be three, such as the recently born human who had a third parent—a clever medical technique for ensuring that the child would not have an inherited disorder.

Given that the parentage of the life cycles of animals and stars is different, we might expect inheritance to work in different ways—and so it does. Some biologists might be tempted to claim that there's no such thing as inheritance for stars, but that's not quite true. If by "inheritance" we mean some effect, initially unspecified, of the "parents" on the "offspring," then there certainly is at least one such effect in the stellar realm, as follows.

A digression is necessary here into a strange habit of astronomers, one that drives chemists crazy. In astronomy, all elements heavier than the lightest two—hydrogen and helium—are collectively referred to as "metals." To you, me, and a professional chemist, this seems odd. This curious usage certainly overlaps with a more normal usage of the term "metal"—iron is a metal from either point of view. But the astronomer's "metal" is much broader than the chemist's one. To an astronomer the oxygen that we breathe is a metal, but to a chemist it's definitely not. Anyhow, the reason we need to know about this is because an interesting feature of stars that can be measured is their *metallicity*. This term is based on the astronomer's "metals"; hence it means the proportion of stellar material that is composed of everything other than hydrogen and helium. Our Sun has a metallicity of about 2 percent. This is higher than the metallicity of the earliest stars, and it's probably lower than the metallicity of the stars of the future. Why?

The earliest stars were born when the universe was very young. At that stage, after the primordial nucleosynthesis that we examined earlier, there was plenty of hydrogen and helium, but very little else, and certainly no oxygen, iron, or gold. So these stars had a metallicity that was very close to zero. But the more stars that have died before any new star forms, the higher the metallicity of that new star will be.

This trend, caused by the effects of parent stars on child stars, is a form of inheritance in the broad sense of that word. But it's very different from the inheritance that operates in humans and other animals. If stellar inheritance was like animal inheritance, then big stars would beget big stars and small would beget small. But it doesn't work like that. The size of a new star is determined by the size of the cloud—or cloud fragment—that collapses to form it, and this is independent of the size of the stars that produced the cloud, which will anyway be so numerous that they'll almost certainly have been a great mix of large, middling, and small.

In a way, the increasing trend in metallicity that's caused by stellar inheritance is like animal *evolution*. For example, it's a bit like the increasing trend in height that has characterized the proto-human lineage leading to us. But this is a superficial similarity because the two trends are driven by very different processes—natural selection on the one hand and nuclear fusion on the other. Because of this, one is reversible while the other is not.

One side branch of human evolution that led to extinction was the one that produced the species known casually as "the hobbit." The official name of the species is *Homo floresiensis,* because these diminutive humans (only about a meter tall) lived on the island of Flores in Indonesia. Although we don't know why natural selection went into reverse in this case, it is famous for doing this rather often, and in some cases we do know the reasons—such as the moth that went from pale to dark and back to pale again as the tree trunks and branches it rested on changed in the same way due to changing amounts and types of pollution. As far as we know, there are no stars in which nuclear fusion has gone into reverse.

Another difference between animal and star life cycles is that in the case of stars we can answer the question about the chicken and the egg. Atoms and clouds of gas came before stars. We'll come back to this issue later. But for now we turn to what can be considered to be the start of an animal's life cycle, inasmuch as it has a start—the egg.

Chapter Six

The Moment of Conception

A Timescale for Your Origin

Unlike the beginning of a star, which takes many thousands of years, the origin of you, me, or another human takes a very short time indeed—but how short is "short"? I've described it as a *moment,* but that description involves a bit of poetic license. Let's go beyond poetry and examine the science of making a human. This involves the penetration of an egg by a sperm, and the formation inside the egg of the first nucleus containing the combined (maternal and paternal) genetic material for the new individual. We'll now dig deeper into the mysteries of how you or I came to be.

The "moment" of your conception lasted for about a day, give or take. If the sexual act that led to *you* took place sometime on Monday, you had probably originated by the end of Tuesday, and certainly by the end of the week. The primordial you was only the start of a very long developmental journey; one that took, depending on how you look at it, nine months, fourteen years, or decades. These are the approximate times to birth, sexual maturity, and now, respectively. But here we're concerned just with the start of this incredible journey. That too was a journey, literally speaking, or rather two journeys: one by an egg, the other by sperm.

If you were to stretch a human sperm cell out straight, it would measure about a twentieth of a millimeter (1 / 500 of an inch) from its head to the tip of its long tail. Invisible to the naked eye, but only just so. To make fertilization happen, such a cell must travel the distance from the vagina to the egg, which is more than 10,000 times its own length. That's a bit like you or me running a marathon. It's not a fair comparison, though, for several reasons. An important one is that the female reproductive tract assists sperm cells on their way, while the streets of London, for example, definitely don't assist the runners.

The egg has a journey to make too. Having been produced in the ovary, it travels from there along the Fallopian tube, usually meeting up with sperm cells (if there are any to meet with) in the region of the tube called the ampulla. This is where fertilization typically takes place. So the spatial scale of fertilization is measured in centimeters or inches, the temporal scale in hours or days.

One Human or Two?

We've all met twins. Some of us even *are* twins—well, half of a pair of twins might be a better way of putting it. But we humans most often arrive in the world one at a time. It's worth asking why. However, this is more than one question because there is more than one type of twinning—the types referred to in everyday speech as identical and non-identical. Or, to a biologist, monozygotic and dizygotic, with *zygote* simply meaning "fertilized egg." The ways in which the two types of twins come about are very different from each other, and are hinted at by the biologist's names for them.

Identical twins arise from the fertilization of a single egg by a single sperm, just as does an individual who is not a twin. The difference is that a fertilized egg that goes on to produce twins splits and gives rise to two embryos. The split occurs at an early stage. Thus, if identical twins are to be born, there will be *either* two of the mulberry-stage embryos that we saw in Chapter 3 *or* just the usual one of these, with the split occurring very shortly after. At whichever precise stage the split happens, its result is two early-stage embryos that derive from a single fertilized egg. In my case (and probably in yours; the frequency of splitting is less than 1 percent) there was just one. Well, that's likely to be true, though a splitting into two followed by the death of one of the mulberries might never be detected.

In contrast, non-identical twins arise from the fertilization of two different eggs by two different sperm—entirely independent events. This is only possible when the mother's ovary has "accidentally" released two eggs into the Fallopian tube at the same time—such releases normally occur singly. Clearly, there can be two fertilizations only if there are two eggs. Having two sperm is hardly a problem, since a single ejaculate contains many millions of them.

Double (and multiple) ovulation is a rare event, just as is the splitting of a fertilized egg. Thus humans usually originate one at a time. But quality is important as well as quantity. Suppose that there's just one egg to fertilize and the resultant embryo never splits. This means no twins. However, imagine a single egg fertilized by two or more sperm—a situation called polyspermy. What would happen here? Usually this kind of fertilization is fatal for the would-be person. A mixture of three or more complete sets of genetic material simply doesn't work. This really is a case of two's company, three's a crowd.

But what sort of stuff is this *genetic material*? How does it work? What determines our individuality? How do genes interact with environmental factors? What's the current status of the nature-versus-nurture debate? These are the questions to which we now turn.

The Kingdom of the Genes

The race between millions of sperm is finally won when a single sperm cell enters the egg. At this point, various mechanisms come into play to block other sperm from entering, and thus to prevent (in most cases) the occurrence of polyspermy. But the winner had less than a one-in-a-million chance at the outset. If it had been a horse, even the most courageous of gamblers would have been loath to put money on it winning the race. And although there's usually only one egg at a time, a woman will typically ovulate a few hundred eggs in her lifetime, of which only a few at most may be fertilized. One of the main reasons that you are you and I am me is which egg and which sperm were involved in "our" fertilizations. If an almost-winning sperm is overtaken by another one just before the barriers come down, the (real) individual who is later born is in a strange sense the brother or sister of the (hypothetical) one who would have been born if the overtaking event had not happened. And since we've started thinking about brothers and sisters, albeit in an unusual context, let's continue with the question of what determined your sex. This is as good an entry point as any to the kingdom of the genes.

In the nucleus of almost every cell of your body reside 46 small rod-shaped things called chromosomes. Given appropriate preparations, you can look through an ordinary microscope and see them. They consist of 23 pairs, but if you're male, one of the pairs (XY) doesn't match as well as the others, whereas if you're female, then the members of your pair of sex chromosomes (XX) match each other just as well as do your two copies of chromosome 5 or of chromosome 21 (for example).

Despite ill-educated kings of the past blaming their queens for giving birth to daughters rather than sons, it was the sperm and not the egg involved in your fertilization that determined your sex. About half of the sperm cells that constituted the ejaculate from which part of you came were "male sperm"; the other half were "female sperm." Of course, that's nonsense because sperm don't have a sex; it's just shorthand for "male-determining" and "female-determining" sperm cells. These are, respectively, those carrying a Y-chromosome and those carrying an X. When one of these is involved in a fertilization with an egg (which is always X), the result is either a male or a female embryo. There are, however, a few unfortunate cases where a mistake occurs in the production of egg or sperm and you get a non-standard zygote such as XXY, which produces an intersex outcome, though one that's closer to male than to female.

By the way, X and Y are just labels, not shapes. Chromosomes don't bifurcate. The X is typically longer than the Y but is of the same general rod shape. And there's nothing magical about having XY as the male and XX as the female—in birds the system is flipped over, so here the egg, not the sperm, determines the sex of the offspring. In some animals, such as turtles, there is no pair of sex chromosomes, and the sex of an embryo is determined not genetically but by the environment—specifically the temperature of the nest.

Now, back to humans and on to the crucial issue of what happens to those rod-shaped chromosomes when cells divide. A cell in your stomach lining typically lasts about a day. Its short life span is hardly surprising since it's bathed in hydrochloric acid. But if these cells died without replacement, we'd have holes in our stomach walls, with dire consequences. Instead, what happens is that nearby stem cells divide to produce daughter cells that replace, in an organized way, the ones that died, so the stomach lining remains intact. If you're female, the new cells (like the old) are all XX; if you're male, they're all XY—not that a stomach cell cares too much about sex.

Most of the cell divisions that occur all over your body during the course of your lifetime are the same as those that produce new stomach cells, in terms of what happens to the chromosomes. But one of the divisions in the cell lineage leading to eggs or sperm is different. Instead of an XX egg progenitor cell giving rise to two XX daughter cells, the XX cell gives rise to two single-X cells. And likewise, a sperm progenitor cell gives rise to a single-X cell and a single-Y cell. In this way, eggs and sperm come to be unique among the cells of our bodies in having only half the amount of genetic material to make a human. This halving is then precisely balanced by the doubling that occurs during fertilization—so we get back to where we started.

The halving and doubling is true not just of the sex chromosomes but of all the other chromosomes too. You and I have two copies of each of these, but our eggs or sperm don't. *The whole set of chromosomes is halved* in the lineages that lead to eggs and sperm, but maintained in all other cases. And which half you get helps to make you who you are.

But in a sense there's no such thing as a fixed "half." It's true that each of us received half of our chromosomes from our father and half from our mother. But those two halves do not remain intact in the production of our own eggs or sperm. Rather, the destinations of, for example, my maternal and paternal copies of chromosome 6 may be the opposite, in terms of which of a pair of my sperm cells they end up in, to the destinations of my equivalent copies of chromosome 7. This is a casual way of stating one of Mendel's laws of inheritance—the law of independent assortment, which proclaims the independence of what happens with any one chromosome pair over what happens with any other. (By the way, if you're interested in finding out more about Gregor Mendel, the monk who founded the science of genetics, there's an excellent book about his life called *A Monk and Two Peas,* by Robin Marantz Henig.)

Not only are maternal and paternal chromosomes reassorted when we make our own sex cells, but bits of them can break off and be recombined with the "wrong" one, thus making new chromosomes that are hybrids between the copies that we inherited. And since the chromosomal rod is in fact a string of beads (genes), if you don't mind a mixed metaphor, this means that two beads that initially sat on opposite sides of a break can end up in different chromosomes. Picture a string of orange beads alongside a string of purple ones as the starting point; in this case, the end point is one string of beads that starts orange and ends purple together with another string that does the opposite.

The string-of-beads analogy is crude, but it captures the essence of the connection between genes and chromosomes. We now know of various complexities such as long gaps between beads and even gaps within beads, but these don't alter the picture in a significant way. For example, the fact that there are far more beads than rods remains true. Since we humans have 23 pairs of chromosomes and about 23,000 genes (a figure that I've "rounded" to make the sum easy), there are about 1,000 beads per rod on average.

Our totality of beads constitutes our *genome*. Each of our 23,000 or so genes does a different job. One makes insulin, for example. Another makes a digestive enzyme called amylase (actually there's a family of these), which is contained in our saliva. And some act in pairs. For example, the hemoglobin in our red blood cells is a collaborative effort between two genes—and they're even on different chromosomes. Our genome, with its thousands of genes, makes thousands of proteins, all of which do particular jobs in various parts of our bodies, including our brains. But we are not just the children of genes. In a sense, our upbringing is a parent too.

Nature versus Nurture

At the moment of conception, each of us has all of our genes. We will never get any more, despite the fact that we eat genes. Our food—whether meat or vegetable—consists of tissue made of cells, within each of which is a complete genome of the animal or plant concerned. Most of their bodies and ours are composed of four types of big molecule—proteins, carbohydrates, fats, and nucleic acids (the DNA of the genes and the RNA through which the genes work). This leads to an interesting question: why do those lists of things our food contains often show the first three of these but never the last? Do we not want to be reminded that all of us, even vegans, eat genes?

But eating genes doesn't add to our own. Depending on how old or how processed our food is, the genes may have already decayed by the time we eat them. And anyway, our digestive system breaks down all large molecules into their constituent smaller ones, so intact genes never get into our bloodstream from our food.

Although we have all our genes at the moment of conception, we have none of our thoughts, and indeed none of our nerve cells that will ultimately produce them. So in a genetic sense all of the prospective individual is there, while in a psychological sense none is. Does our psychology follow inevitably from our genes? Absolutely not. Mental traits have a certain heritability, as do physical ones, but in general their heritabilities are lower than those of their physical counterparts, like height. Our minds are not prisoners of our genes.

One type of study that has led to this conclusion involves comparing identical twins who have, for a variety of reasons, been raised apart—same genes, different environment and upbringing (or, to use other words, same "nature," different nurture, though equating genes with nature is a bit naughty). These studies typically reveal two very different people. And even when identical twins are raised together they often turn out quite different in character, albeit very similar in appearance. I knew such a pair of twins—albeit I was better acquainted with one than the other—when I was a student. One was studying science, and in his spare time got involved in martial arts, especially judo. The other was studying philosophy and in his spare time read and wrote poetry. The message is: genes do not make people. Each of us is a product of our inheritance, our upbringing, and the interaction between the two, which is complex and incompletely understood.

Conception of Other Animals

Ironically, given its name, a cock (or cockerel or rooster) has no penis. In fact, this is true of most male birds—the male blackbirds that fight each other at a territory boundary at one side of my garden have no penises either. So when conception occurs in such bird species it does not involve a copulation, at least in the sense that we understand that term. Instead, the male and female birds get involved in what's called a cloacal kiss. This is where the two birds briefly make contact at their cloacas, which are combined openings for reproduction, excretion (urine), and egestion (feces). During contact, the male's cloaca transfers sperm to the female's. The sperm moves into the female's reproductive tract and is later used to fertilize eggs—"later" in one sense only, the sense of being a considerable time after sperm transfer. In another sense it has to be quite early—a bird's egg as we see it would be impossible for a sperm to fertilize. So the moment of conception for a bird is well before the egg is finalized—let alone laid—and in particular before it comes to have a hard shell.

Apart from the lack of a penis, birds are much like mammals in their conception. In both cases the new individual comes into existence within its mother—internal fertilization. But many marine animals use a system of external fertilization, in which the sperm and eggs are released into the surrounding seawater, where lots of fertilizations of eggs by sperm take place. This has been much studied in sea urchins—those spiny balls that are closely related to starfish. But it's not only invertebrates that have external fertilization. Marine vertebrates generally use the same technique. Most fish simply shed their eggs and sperm into the sea, "hoping" that there will be many a successful rendezvous between the eggs and sperm concerned. There are a few exceptions, though—for example, in shark species copulation occurs and fertilization is internal.

The living world is full of exceptions. For any taxonomic group you choose, attempts at generalizations usually fail. As we've seen, most male birds don't have penises, but in a few species they do. Also, most fish have external fertilization, but a few have its internal equivalent. Indeed, the living world contains so many species and so much variation that even the categories about which we try to generalize may not be as distinct as we first think. Take centipedes, for example. Do most of these elongate animals have internal or external fertilization? Well, it's internal, but only after the sperm has had a period of being external, contained in a sac that's suspended in a web of silk-like strands attached to lumps of soil or other environmental prominences. Having deposited its sperm thus, the male centipede wanders off, leaving the female to come along and find it, take in the sperm, and use it to fertilize eggs sometime later. So, unlike the situation in mammals, where the sperm goes straight from the male to the female, the sperm goes from the male to the female via an environmental interlude.

What's in a Name?

You'll probably have noticed that so far in this chapter I've used the words *conception* and *fertilization* more or less interchangeably. But is it correct to do so? Therein lies a can of worms. Can we really speak of the conception of a centipede or a snail? Or is this a misuse of a purely human term? When a biologist rather than a priest talks or writes about human reproduction, *fertilization* is normally used rather than *conception*. Again, is this okay? As always in relation to such matters, I consulted my Concise Oxford Dictionary. In the entry for *conception* it refers to the origin of a child, while in the entry for *fertilization* it refers to an animal or plant. This suggests the two have different usages, though I'm only guessing that the dictionary intends *animal* to be used in the sense of what a biologist would call a "non-human animal."

Interestingly, if you type *conception* into Wikipedia it redirects you to *fertilization,* hence perhaps implying synonymy. My third source was even more interesting. My "bible" of developmental biology is a book of that title written by the American biologist Scott Gilbert. If you look up conception in the index of the latest edition of this book, it's not there. Not even so much as a "see fertilization." If you look up fertilization, on the other hand, you get a substantial block of entries. But on the first page of the relevant chapter, "fertilization" and "conception" are used interchangeably.

It looks like we're confused about this issue. But probably this confusion is just a hangover from history. Probably, if we think about it carefully, we're not really confused at all. Once upon a time, at least in the Christian world, the conception of a human was thought to be an altogether different thing from the fertilization of the egg of a blackbird or a dog. One involved an immortal soul, while the other did not. Humans were not animals. Now, however, it would be appalling arrogance to try to make such a clear-cut distinction. Well, maybe not in some brainwashed sects, but definitely in the world of educated, broad-minded people. The coming together of an egg and a sperm to create a new life is a process we share with our mammalian, and more distant, cousins. We can call this process anything we like.

III

IN THE BEGINNING

Chapter Seven

A Universe Begins

Quicker than Conception

On the scale of the universe, a human is not just a tiny speck; each of us is a speck within a speck within a speck. The spatial scales of the familiar everyday world and the part of the universe that we can see—the observable universe—are so different that words fail to convey the magnitude of the contrast. And yet, vast as it is, the origin of the universe took less time than the origin of you or me.

That's a strong statement and it probably needs some defense. So here goes. We saw in Chapter 6 that human conception takes about a day. That's the typical time from having sex to the fusing of the genetic material of the egg and sperm to produce the first nucleus of a new human. Both the start and the end of this period of time are well-defined events, so it's easy to put a figure to the period in between—albeit with a bit of a wobble, as the time it takes the winning sperm to reach the egg is quite variable.

In the "conception" of the universe, the initial event that we call the big bang was well defined and almost instantaneous, even though it's hard to envisage. One way of characterizing it would be to call it the start of time. But there's no event in the conception of the universe that's equivalent to the fusing of two sets of genes that defines the end of the conception of a person. Given that, when do we say that the universe's origin was over and its routine existence had begun? Well, there are several possible choices, and we should realize that whichever point we choose for the "end of the beginning," it's simply a helpful marker to enable us to reach a better understanding. The evolution of the universe is like embryogenesis in that it's a continuous sequence that we can break up into stages for our convenience, but we must acknowledge the difference between convenience and reality.

With that caveat, I'm going to choose the creation of the basic building blocks of atoms—protons, neutrons, and electrons—as the end of the beginning of the universe. With this choice, the universe's origin, or conception, lasted approximately one second. It seems bizarre that the building blocks of everything could be made in such a short period of time, and yet as far as we know, this is what happened. Let's now dissect this brief time span.

The First Millionth of a Second

One common misconception of the big bang is that it involved the explosion of an enormous rock sitting somewhere in space, which blasted it into tiny fragments, all of which, a fraction of a second later, were hurtling away—both from the initial site of the bang and from each other. Although that would fit with what astronomers see—lots of galaxies hurtling away from each other and from us—it's the cartoon version of the ultimate beginning, and it's very different from the real one. Science's current view of the real one is even stranger than the cartoon. It's still tentative, but this is what we think . . .

About 13.8 billion years ago, there was nothing. Not just no matter or energy, but no space or time either. All four of these familiar things were yet to come into existence. They did so almost instantaneously through something called a singularity. Picture it not as a large lump of exploding rock in space, but rather as a tiny pinprick in nothingness. Since space is black, we have to picture a spaceless nothingness differently. White seems a poor choice since it represents light. Personally I picture it as gray, though a metaphorical gray. But if metaphorical turquoise or orange works for you, that's fine.

So the beginning was small, not big, and "before" the explosion everything (in other words, nothing) was gray, not black. And there's another difference between the way we believe the universe originated and the cartoon view. After an ordinary explosion, such as that of a bomb, the speed at which the fragments hurtle away is initially fast, but there is rapid deceleration and soon enough all motion has ceased. In contrast, the expansion of the universe appears to have accelerated fantastically a tiny fraction of a second after the bang. We call this cosmic inflation. We're pretty sure it happened, yet we're clueless as to its cause. After inflation, the expansion slowed down, but today it's again accelerating, though nowhere near as much as during the era of inflation. You can't get this crazy pattern from an ordinary explosion.

At the end of the inflationary era—and we're still well within the first *trillionth* of a second—both matter and anti-matter came into existence in the form of quarks. Here's another scientific word that has made the leap into everyday speech. A *quark* is a subatomic particle of which the more familiar particles called protons and neutrons are made—three quarks each. But quark is also a type of cottage cheese that I buy in the supermarket quite often. The universe at the end of inflation can be thought of as an ocean of quarks, with smaller particles, such as electrons, floating around too. And there were anti-matter versions of all of these, as well as their "normal matter" versions.

When the universe was about a billionth of a second in age, the quarks and anti-quarks got together in threes to form protons, anti-protons, neutrons, and anti-neutrons. Then, at about a millionth of a second, the matter and anti-matter particles annihilated each other. If the amounts of the two had been exactly equal, this would have destroyed all of these particles. But they weren't. It seems that there was a tiny excess of matter over anti-matter, with the result that for every billion annihilations one proton or neutron survived. This tiny excess made the difference between galaxies, stars, planets, and life, on the one hand, and—well—none of these, on the other. Quite a weird thought.

An oft-used saying that facts are stranger than fiction is appropriate here: *you couldn't make this stuff up*. The current model of the origin of the universe is utterly bizarre. But we've only seen part of it. Let's now carefully continue our journey into the bizarre realm of the early universe.

The Clock Strikes One

Not one hour but one second. At this point in time, electrons and anti-electrons annihilated each other much as their bigger cousins had done earlier. Again, there was a slight excess of electrons over their anti-matter equivalents at the start of the process, so at the end there was about one electron left for every billion annihilations. We think that the imbalance between electrons and anti-electrons was the same as that between protons and anti-protons, with the result that the universe ended up electrically neutral.

At this point all the free quarks were gone, and so was the anti-matter. The universe was composed of lots of nice normal things—the protons, neutrons, and electrons that we learned about in school. Since these are all subatomic particles on their own, we didn't yet have any atoms. But, strangely, we did have some atomic nuclei. This is because a *hydrogen nucleus* and a *proton* are the very same thing. The nucleus of hydrogen, the simplest of all elements, consists of a single proton and no neutrons—or a single red ping-pong ball and no green ones, using the visual model developed in Chapter 4. So effectively there was an awful lot of hydrogen, but no nuclei belonging to any other elements, when the universe was one second old.

This is what I've been calling the *end of the beginning*. However, bearing in mind the earlier caveat that the evolution of the universe is a continuous process with no breaks between "stages," just as is the case with the development of embryos, we need to look at what happened next. But we've already done this in Chapter 4, so all that's necessary here is a quick reminder of the process involved.

For about half an hour the process of primordial nucleosynthesis took place, through which a large number of the hydrogen nuclei were converted into helium nuclei by nuclear fusion. Theoretical predictions of the resultant fraction of helium (by mass) in an overall medium consisting largely of hydrogen produce a value of about a quarter. This should correspond to the fraction of helium found today in regions of interstellar gas that have not yet been enriched with debris from dead stars—and it does. This is a remarkable result, and one that is taken as strong evidence that the big bang theory is correct.

But why should primordial nucleosynthesis have stopped? If there was still plenty of fuel (hydrogen nuclei), which there was, why didn't it just keep getting burned—in other words, converted into more helium? Why did hydrogen continue to be the most abundant element, rather than a comparatively rare one? The answer is temperature. Nuclear fusion requires a very high temperature in order to proceed. There were claims in the 1980s by a pair of scientists, Martin Fleischmann and Stanley Pons, that they had succeeded in achieving "cold fusion," but subsequent attempts by others to replicate this result showed that it was almost certainly wrong.

The early universe started extremely hot and then rapidly cooled. Because of this, our current model of the origin of the universe is often called the *hot* big bang model. How hot is hot? At the beginning it was many trillions of degrees. But by the time the clock struck one, it had dropped to a mere 10 billion degrees. That's hot enough for nuclear fusion, but the temperature half an hour later (less than 1 billion degrees) was not. So, fusion of hydrogen into helium ceased. It only restarted much later, and then only in the hot centers of stars. But at this point—half an hour after the big bang—there weren't any stars yet. There was just a vast ocean of hydrogen and helium nuclei, electrons, and a lot of light, but not much else—except, perhaps, for those hypothetical particles that are thought to make up the mysterious stuff we call *dark matter*.

The Long Wait for Atoms

So far in the story, most of the important events in the early universe took place in the first second, with primordial nucleosynthesis being the slowpoke and taking up to half an hour. But when nuclear fusion ceased there was a long interlude in which not much happened—at least not abruptly. For the next quarter of a million years the universe continued to expand and cool. Since at an age of half an hour it had already been cool enough for nuclear reactions to stop, the further cooling had no effect—nuclear reactions simply continued to be impossible. But the further cooling and expansion that occurred between a quarter million and a half million years after the big bang took the universe across a very different threshold.

Remember that up to this point there were lots of atomic nuclei, mostly those of hydrogen and helium, and lots of free electrons, but no atoms—in other words, the nuclei and electrons didn't combine with each other. There was also a lot of light in the form of the light packets that we call photons (brought into the public's imagination by Star Trek's *photon torpedoes*). A photon is usually drawn as a little squiggly arrow. This is to indicate that it's a kind of hybrid between a wave (the squiggle) and a particle (its short length).

One way of thinking about the early universe is that it was foggy. What this means is that if you imagine yourself to be a photon of light, you would keep bouncing off free electrons and nuclei. So you would never get very far in any direction before you hit an obstacle and ricocheted off it in a different direction. Your path would be like a drunkard's walk rather than a sprinter's 100-meter dash.

Then, as the universe continued to expand and the temperature fell to just a few thousand degrees, protons and electrons combined to produce hydrogen atoms. Since a hydrogen atom consists of only one of each, this reaction can be written down in a very simply way: proton + electron → H + photon. The letter H stands for a hydrogen atom—the eye in the illustration at the start of this chapter (page 72) is based on a photograph of such an atom. But what's a photon? We've already seen that it's a tiny package of light whose nature is part wave, part particle. However, different wavelengths of photon have very different properties: some are lethal to humans, others are neutral, and others again are helpful—at least in the right doses.

We're all familiar with X-rays, and in particular with X-ray photos of decaying teeth, fractured bones, and other medical conditions. X-rays are beams of radiation that are closely related to beams of visible light—the difference is that they have a shorter wavelength. Our eyes can see beams of light as long as their wavelengths are between certain values—those that correspond to the colors red (longer) and violet (shorter). Wavelengths shorter than violet are called ultraviolet, and we know very well that these exist and are dangerous—we can prove it by going from an Irish winter to an Italian summer and lying on the beach all day.

As the wavelength gets even shorter, the danger gets greater. Shorter than ultraviolet takes us to X-rays. Half an hour on an Italian beach would be fine for a northern European, but half an hour in an X-ray machine would not. And we're now approaching gamma rays. They have wavelengths even shorter than X-rays. So, from a human point of view, they are the most dangerous of all.

As ever, categories are useful to help us understand things. But as we've already seen, their boundaries are often fake. A human embryo does not pause at a certain point to say, "I'm just leaving the mulberry stage now." And there is no signpost on a beam of radiation saying, "I'm at the boundary between X-ray and gamma-ray wavelengths." These two categories, plus ultraviolet and visible light, plus those beams that have even longer wavelengths, such as "microwaves" and radio waves, collectively make up what we call the electromagnetic spectrum of radiation—a concept that we owe to the nineteenth-century Scottish physicist James Clerk Maxwell.

The reason for using quote marks in the case of "microwaves" is that in the grand scheme of things their wavelengths are long, not short. So the "micro" is a misnomer, at least from our human reference point of visible light—the wavelengths that we can see. Microwaves have longer wavelengths than such light, not shorter. So, contrary to some non-scientists' concerns, they are not dangerous—at least no more so than the wavelengths that carry radio and television broadcasts, and those are generally reckoned to be benign.

Let's get back to the early universe after our digression into the nature of visible light and its electromagnetic cousins such as gamma rays. When the universe was a little under 0.4 million years of age (still very young compared with its current age of 13.8 billion), the "recombination" of protons and electrons into atoms occurred, although, like "microwave," this is a misnomer. As far as we know, before this time protons and electrons had gone their separate ways ever since their formation. So this was their *first* combination, not a *re*combination, despite the latter being the commonly used term.

Anyhow, whatever we call it, this process happened all over the place—right through the universe. So, everywhere, atoms were forming all at once. At the end of this process there were few free electrons and nuclei, but lots and lots of atoms. Since atoms are the building blocks of matter, their formation was a crucial step in the continuing development of *structures* in the universe, be they stars, planets, or people. However, this phase in the history of the universe was important for another reason too. To get an idea of this, we need to become photons again—or to magically shrink ourselves and sit astride them.

When we pretended to be photons before, the universe was younger and we kept bumping into things, particularly free electrons. Now these obstacles have been removed—they've become bound up with nuclei to form atoms. So the miniature you and I, sitting astride our squiggly photons, can dash across the universe relatively unhindered. We're no longer drunkards or even sprinters; we're long-distance runners of unbelievable speed, whose mega-marathon-scale track is remarkably straight.

Seeing the Past

When the big bang theory was being developed in the mid-twentieth century, the people involved thought that if it were true then we should be able to see the radiation that was last scattered by free electrons less than 400,000 years after the bang. But what type of radiation would this be? Recall that when protons "mated" with electrons to form hydrogen atoms, the radiation given off was in the form of photons with short wavelengths—most of them were in the visible part of the electromagnetic spectrum. But they wouldn't have stayed that way for long. As the universe expands, space stretches. Think of the universe as a sheet of black rubber being pulled apart at its four corners and you'll have a 2-D model of what happened, and is still happening, in *three* spatial dimensions. This is another way that the real big bang differs from its cartoon counterpart. And, as space stretches, so do waves traveling through it. So their wavelengths get longer. Although copious amounts of short-wavelength photons were produced when atoms first formed, these are not so short now. In our present, nearly 13.8 billion years later, they would have been stretched so much that they'd have become microwaves.

So the pioneers of the big bang theory in the mid-twentieth century thought that there should be microwave radiation coming at us from all directions, a relic of the last scattering of photons by the fog of free electrons and nuclei. Thus the future discovery of such radiation became a goal—if it was found, the theory might well be true; if it wasn't, then the theory was probably false. Strangely, the cosmic microwave background radiation, as we now call it, was discovered by accident in the 1960s. The two astronomers who made the discovery—Arno Penzias and Robert Wilson—were awarded the Nobel Prize for physics.

Penzias was born in Germany, Wilson in Texas. Years later, their individual tracks through time and space came together in New Jersey. They were working with a device called the Horn Antenna located in the township of Holmdel. This device was a large horn-shaped metal structure intended for radio astronomy. In other words, it could be pointed at any area of the sky, much like an ordinary telescope, but its job was to see not light waves but radio waves. To do this job well, the device must be working properly, and one aspect of this is that its users should see only the signal—the radio waves they're looking for. This signal should be uncontaminated by *noise* in the sense of background radiation. But Penzias and Wilson found such noise, and moreover, its nature and amount seemed to be independent of the direction in which they were looking. They tried all manner of things to get rid of it, most famously cleaning pigeon droppings out of the antenna. But still it was there.

The two astronomers consulted various colleagues to try to solve the problem. One of these was the physicist Robert Dicke, just down the road at Princeton University. Dicke was a proponent of the big bang theory and was one of the scientists who were keen to discover the cosmic microwave background radiation that their theory predicted. In fact, he was in the process of building a detector to try to find it when Penzias and Wilson came to him with their "problem." The rest, as they say, is history. The annoying noise in the horn antenna was the hoped-for cosmic radiation. The apparent problem was really a solution—a solution to the debate between supporters of the rival big-bang and steady-state models of the universe. Victory for the big bang.

An Event without a Cause?

But what sort of a victory? The big bang theory is unlike a normal scientific theory for a very specific reason. Most scientific theories try to explain an event in terms of a cause, or perhaps a combination of causes. Theories about the origin of life on Earth (Chapter 8) try to explain this origin in terms of chemical processes that took place on the early Earth and led to the first life-form. But although the big bang theory is an attempt to explain the origin of our universe, it does not provide us with a cause. This state of affairs is scientifically unsatisfactory but, as far as we know right now, inevitable, for the following reason. Logically, a cause must precede its effect. You cannot propel a stone across a river by throwing it after it has already landed on the far bank. This logical sequence of cause first, effect later, is so familiar to us in everyday life that we don't often think about it.

Unfortunately, the big bang was the event that led to the existence of *time* as well as space, matter, and energy. Which means that time started with the big bang. If that's true, as is currently thought, there was no time before the big bang in which a cause could have operated. Hence the origin of the universe in this manner is sometimes described as an event without a cause. The big bang theory is a causeless model of how our universe began.

From the perspective of another area of science (such as biology) or of everyday life, this is crazy. And yet it's where things stand at present. Since it seems like all this is becoming science fiction, let's now turn to that realm. The English sci-fi writer Douglas Adams said in *The Hitchhiker's Guide to the Galaxy:* "There is a theory which states that if ever anyone discovers exactly what the Universe is for and why it is here, it will instantly disappear and be replaced by something even more bizarre and inexplicable. There is another theory which states that this has already happened." Perhaps if someone someday improves upon the big bang theory and devises a theory that explains how everything *really* got here—in terms of a cause—an "Adams replacement" will immediately occur and we'll all disappear.

Of course, when faced with a situation such as this—a universe coming into existence almost instantaneously from a "preceding" state of nothingness—we all look for a more satisfying explanation. Some people seek this in religion, some in a broader scientific perspective, such as a multiverse—the idea that our universe is just one out of many. But maybe we're all seeking the impossible. Maybe a close cousin of the chimpanzee who evolved to survive in Africa a long time ago shouldn't expect to be able to understand the origins of everything. Maybe, but that won't stop us from trying.

From Atoms to Aliens

The long wait for atoms—about 400,000 years—was doubtless a tiny wait compared to the wait for life. Suppose for a moment that Earth was the first planet to evolve living creatures. In that case, the life-wait was from the universe being a mere infant at 400,000 years of age until it had reached an age of about 10 billion years (its current age of 13.8 billion years minus the 3.8 billion years of life on Earth; Chapter 8 focuses on that second figure).

But this is an unlikely scenario. We thought we were the center of the universe; now we know better. We might have been the first planet to evolve life, but this seems just as improbable as us being at the center of things. If we're not spatially special, why should we be temporally special? Indeed, the first planet to evolve life was probably not even in our Milky Way galaxy. The odds are firmly against it, given the billions of galaxies that exist.

A more likely scenario is that Earth is a middling planet in the race toward life. There are probably planets somewhere out there that are far ahead of us and others that are far behind in this respect. The trouble is that we have no way of estimating *how* far. There are too many imponderables. But it's entirely possible that there was alien life somewhere when the universe was 5 billion years old rather than 10 billion. Maybe at that early stage it only took the form of alien microbes. But if an evolutionary process similar to ours started with those microbes, by now there may be intelligent aliens on the planet concerned that are vastly more advanced than us. This conclusion assumes that intelligent life can survive the first few centuries of having nuclear weapons without causing its own extinction—an issue that we'll come back to later.

Chapter Eight

The Opposite of a Whimper

Infant Earth

The origin of life on Earth was a much slower process than the origin of the universe in the hot big bang. Nevertheless, the oldest fossils show that life began here surprisingly soon. The earliest life-forms that we know of from fossil evidence were single-celled creatures, not too different from today's bacteria.

Single cells are, of course, very small. But some bacteria can build structures that are rather large. These boulder-like things can still be found today, most famously in the World Heritage Site of Shark Bay, Western Australia. They're big enough to sit on. They consist of many layers, each a kind of thick microbial mat. They're called stromatolites. The very first stromatolites were smaller than this, but by no means microscopic. We know of fossil stromatolites from 3.7 billion years ago that were up to about 5 centimeters (2 inches) in depth. Exactly which group of microbes produced them is not yet clear, but the evidence that their origin was biogenic is persuasive.

A figure like 3.7 billion years ago doesn't mean much unless we can put it in context. So let's do that. Here's a very rough chronology of the infant Earth. Our planet—along with the rest of the solar system—formed about 4.5 billion years ago. For at least the first 0.1 billion years it was a fireball. No incipient life-forms would have stood a chance of survival. This was very definitely an inorganic place (though exactly what *organic* means is not a simple matter; we'll examine this shortly). The first rocks probably solidified just after the fireball era—the oldest terrestrial rocks that have been dated are about 4.3 billion years old. The earliest microbial fossils, as we've just seen, date from about 3.7 billion years. But, as every paleontologist knows, the earliest fossil that has been discovered for a group of organisms only sets the *latest* possible time of their origin. Almost always that origin turns out to have been considerably earlier. Whether, in this case, "considerably" means 0.1 or 0.5 billion years earlier is a moot point. I'm going to take the former view here; time will tell whether this is the correct choice.

So, suppose that the very first single-celled creatures appeared on Earth about 3.8 billion years ago. That's just 0.5 billion years after the first rocks. The brevity of this window of time in which life originated has led some people to propose that life in fact did not originate here, but rather arrived here ready-made from somewhere else. This is the panspermia hypothesis that we touched on very briefly in Chapter 1; one of the main advocates of this hypothesis was the Swedish scientist Svante Arrhenius, who set out a detailed form of it in the early 1900s. There's just one problem with his hypothesis: it's almost certainly wrong.

Not from Space

One of the core principles of all science—from cosmology to biology—is the refusal to accept an unnecessarily complex hypothesis if there's a simpler one that will explain the same facts. In some scientific contexts this is called the principle of parsimony. For example, we believe that the same process—natural selection—has been important in the evolution of all life-forms, whether bacteria, worms, whales, trees, or people. A more complex hypothesis that required us to believe in a different process for the evolution of each group of creatures would be regarded as inferior. Of course, sometimes there is no simple hypothesis that will explain all of the relevant facts, and in that case we have no option but to consider more complex ones. But this is a last resort.

So why have I labeled the panspermia hypothesis as almost certainly wrong? Well, one reason has to do with its unnecessary complexity. Advocates of an extraterrestrial origin of life on Earth replace a single problem with two. The first problem is how living cells arise from non-living matter. We have to deal with that problem either way; whether it happened here or elsewhere, we want to understand the mechanisms at work. But if living cells originated on another planet, we also have to understand how they managed to survive a long journey through space. The fact that some life-forms can be sent into Earth orbit in specially constructed containers that expose them to the vacuum and radiation of space, and survive—which is true for the tiny invertebrates called water bears—is not a license for life-forms to travel intact for much longer times and distances. Although it's not so often emphasized, most of the orbiting water bears died. And it seems likely that the death rate of these or any other space-traveling creatures increases with the time spent in that inhospitable environment. How long would it take for living cells of ancient bacteria from Mars, now long perished on their home planet, to have made their way to Earth? Not days but years. And such bacteria are probably fictional anyhow—we have no evidence of past or present life on Mars, despite the sterling efforts of the *Curiosity* rover and its predecessors such as *Viking 1* and *Viking 2*. If there was never any life on the other seven planets of our solar system, then space-traveling cells would have to have come from *much* further afield. Their journeys would have taken not a few years but thousands or millions of years. The probability of a life-form surviving that much time in space and then waking up and reproducing here on Earth is effectively zero.

There's another reason I don't believe that life on Earth originated from space-traveling cells, seeds, or spores. The idea that 0.5 billion years is a short time in which to evolve life on Earth is misguided. Although this period of time (from 4.3 to 3.8 billion years ago) is only about 1/10 of the Earth's history to date, it's really not short in evolutionary terms. It sounds longer if we convert it to millions—in 500 million years of evolution an awful lot can happen. Take tetrapods (the land vertebrates), for example. Five hundred million years ago, there weren't any. Now there are about 30,000 species of them, ranging in size from the diminutive dwarf gecko and bumblebee bat to the gargantuan blue whale, with us somewhere in the middle of this vast range of body size.

However, we need to be careful with our comparisons. It could be argued that the very first tetrapod was already a highly evolved creature. Not only did it have cells, but it also had organs—pretty much all of those that would later be elaborated to make us and our extant mammalian cousins. So it didn't have to invent any of these things. Perhaps evolution can achieve a lot in a few hundred million years if almost all the components needed for the descendants were already present in the ancestor. But perhaps the origin of life was fundamentally different. Here, transitions had to be made that were later taken for granted—for example, inorganic to organic, and acellular to cellular. Perhaps these are harder transitions to make than later ones such as reptile to mammal.

What Does "Organic" Mean?

Recently I read a news headline that said "Complex Organic Molecules Found in Martian Meteorites." As I read on, the author of the piece rapidly led the readers to a view that similar meteorites arriving nearly 4 billion years ago brought life to Earth from Mars, even though there's no evidence that there ever has been life on the red planet. From the undefined to the ultra-speculative. Science? Definitely not.

We shouldn't get excited by the word *organic*. It has no overall consensus definition. It means different things to different people. Let's look at some of these things, but without doing any real chemistry (which, as you'll recall, I'm very bad at). The only element that is truly crucial here is carbon. All organic molecules contain carbon, while the majority of inorganic ones don't—but even at this early stage of the story things are becoming fuzzy, because some inorganic molecules *do* contain carbon. How so?

A common definition of an organic molecule is as follows: a molecule that has at least one bond between carbon and hydrogen. If we accept that definition, then carbon dioxide, which contains no hydrogen, is not organic. But methane gas, which can be produced by both biotic means (most famously cows farting) and abiotic ones, definitely *is* organic, given that a molecule of this consists of a carbon linked to four hydrogens—that is, it contains four C-H bonds.

Organic can be defined in both more and less restrictive ways. The most liberal definition is "anything that contains carbon is organic." This has the advantage of cleanness—carbon dioxide is now organic. But so is a diamond, which to some people might seem odd. Tighter definitions are that an organic molecule must contain at least two carbons, or a ring or long chain of carbons. With any of these, methane is inorganic, which seems counterintuitive. Anyhow, that's enough ways of defining organic to see the point that this is a slippery word. And it becomes even more slippery in the context of "organic" farming.

So the exclamation "Organic molecules discovered in meteorites!" is meaningless without some clarification of what the exclaimer means by *organic*. Does "*complex* organic molecules" make the discovery more or less clear? Probably the latter, because the word *complex* is usually undefined too. How complex is complex? At one end of the spectrum, we can say that methane, with its single carbon, is fairly simple. At the other end, a DNA molecule, with almost countless carbons, is very complex indeed. But what lies between? And is there a threshold between simple and complex? The answer to the second of these questions is no; the answer to the first is a bit longer.

Since I'm not much of a chemist, I'd like to keep things easy. Let's imagine four levels of complexity of organic molecules: those with (1) one carbon, (2) two carbons, (3) a ring or short chain of a few carbons, and (4) a long linear chain or other shape with multiple carbons. Methane is an example of the first level of complexity. Ethanol, as found in alcoholic beverages, is an example of the second (it has two carbons). Sugars, such as glucose, exemplify the third level of complexity, as do amino acids, which are the building blocks of proteins. Proteins themselves, along with DNA, RNA, carbohydrates, and fats, are all macromolecules and so are examples of the fourth. What now becomes interesting is which level of complexity of organic molecules can be found in meteorites—both those from Mars and those from further afield. And the question that logically follows from that is: does this mean that life came to Earth in this way?

Carbon-containing molecules from the first three levels can be found in meteorites. However, this does *not* mean that life came here from another world. None of the molecules concerned constitutes a life-form. Even the macromolecules of level four, like DNA, do not constitute life-forms. So we can say that the atoms and molecules required for life came from space, but life itself did not. And this is hardly a surprising conclusion. We've already seen, in Chapter 4, that dying stars are the ultimate source of carbon. In their dying they shed this element into space. And if we analyze the stuff between stars—the dilute gas that we call the interstellar medium—we find that it has not just carbon atoms but carbon-containing molecules too. Methane is out there, as is alcohol. There's enough alcohol in the interstellar space within our Milky Way galaxy for all 7 billion humans to get very drunk indeed.

The Primordial Soup and the Primitive Pizza

The transition from inorganic to organic is easy. It has happened countless times, all over the universe. Organic molecules, even quite big ones, are everywhere. But to have the macromolecules of life is another matter altogether, and to have a simple cell like a bacterium, containing lots of these, is something else again. These comparatively *hard* transitions almost certainly took place on the infant Earth, from a starting point of organic stardust. They were the route from stardust to us and our fellow terrestrial creatures. We don't know *exactly* how the origin of life occurred, but we've got some pretty good ideas on the subject, and we can think of the process as taking three steps.

Step I

From the simple organic molecules of levels one and two, the more complex ones of level three (like amino acids) were made in a murky liquid that we call the primordial soup. This covered parts of the surface of the infant Earth, and it contained water, methane, ammonia, and lots of other simple molecules, both organic and inorganic. Suppose a bolt of lightning electrifies such a soup—what happens then?

An experiment conducted in 1952, the year in which I was born, provides an interesting answer to this question. It was conducted at the University of Chicago by Stanley Miller, working under the supervision of Harold Urey, so it's often called the Miller-Urey experiment. Miller had a large glass vessel containing a liquid with some primordial soup ingredients in it, and he heated it so that it became a mixture of liquid and gas. Then he passed an electric charge through it to simulate lightning on the early Earth. About a week later, the mixture was analyzed; much later, further analyses were conducted. The molecules produced included most of the amino acids of which the proteins of our bodies are composed. Our hair, like the fur of our mammalian cousins, is made up largely of a protein called keratin. Miller's experiment produced no hair (that would have been too much to ask), but it produced the amino acid units of which hair is made.

Step II

Step I was rather easy, but step II is much less so. This is to get from amino acids to protein, or to make the equivalent step for the other types of macromolecule. What is needed here is to connect the small molecules up in a certain way, so that they link into a long chain, sometimes in the hundreds or even thousands. Normally this requires enzymes to catalyze the linking-up. But most enzymes are long-chain protein molecules, so we find ourselves in a Catch-22 situation. It's not possible to make DNA without an enzyme. And normally it's not possible to make an enzyme without DNA.

But there's a way out of this particular Catch-22, and it's called RNA. This is a close relative of DNA, as you might guess. They're both nucleic acids, but there are two main differences between them. The sugars that make up the backbones of the molecules are different—ribose for RNA, deoxyribose for DNA. And RNA usually exists in single-strand form, while DNA is in the famous double helix—two long strands twisted around each other.

Nowadays, enzymes are usually proteins; genes are usually made of DNA. Or, to put it another way, proteins do the work, whereas DNA is a store of information. But it turns out that RNA can do both. And perhaps that's where life started—with RNA catalyzing the production of more RNA, possibly not in the primordial soup but rather on a semi-solid surface that the biologists John Maynard Smith and Eörs Szathmáry called "the primitive pizza" in their book *The*

Origins of Life. But even if we accept that RNA once made more RNA in this manner—the RNA world hypothesis—the reproduction of a big molecule is not the same as the reproduction of a cell.

Step III

This is the hardest of all, and in fact consists of plenty of substeps. From a starting point of reproducing macromolecules we progress to reproducing *collections* of macromolecules, and then to more coherent collections that are *bounded*, probably by a membrane not dissimilar to those of present-day cells. These transitions were powered by Darwinian selection. Just as happens with later animals and plants, those tiny primordial organisms that reproduced more effectively than others will have left more offspring.

We now have terms that are routinely applied to life but not to non-life. We have parents and offspring, concepts that would not have been very meaningful in the earlier stages—such as those captured by the Miller-Urey experiment. This doesn't mean that we can *define* life—it's notoriously hard to do so—but we do now have entities with reproduction, variation, and inheritance, albeit of a primitive kind, and those are the characteristics needed by a population of entities for natural selection to get to work on them. From the first microbes to us, natural selection has been working without a break for nearly 4 billion years.

How Many Origins?

There's nothing to stop the process described above from happening twice or even many times. And yet we believe that all present-day organisms on Earth, including animals, plants, and microbes, share a common living forebear—called LUCA, for *last universal common ancestor*. We can think of LUCA as being a sort of protobacterial organism. Of course, scientists don't just believe things for the fun of it. There is always at least some evidence, however inconclusive it may on some occasions be, supporting a belief. And that's just as true about hypotheses of the origin of life as it is of others.

One of the key pieces of evidence in favor of a single origin of life is the universality of the genetic code among present-day creatures. Since this is the first time I've mentioned the code, let's be clear about what it is. It's the way in which sequences of building blocks along the length of a DNA molecule specify equivalent sequences in RNA and ultimately in protein. The code works in threes, so a particular triplet of adjacent blocks within DNA / RNA will always build a single amino acid into a protein. The code is often referred to as redundant, because some amino acids can be coded for by more than one triplet. This redundancy ties in with the fact that there are 64 triplets but only 20 amino acids. Nevertheless, even the redundancy is consistent: it's the same in all species.

This suggests very strongly that all today's organisms are descended from the creature that we call LUCA. If they had resulted from different origins of life, the chances of having ended up with exactly the same genetic code are negligible. But this conclusion does not mean that there only ever was a single origin of life on our planet. There may well have been several. However, if that's true, then it looks like those creatures who descended from LUCA, including us, prevailed over the descendants of other, doomed ancestors, who, perhaps not long after they began their own multigenerational routes into the realm of life, had their lineages extinguished, either through inferiority or by chance. We can never know if this happened, but it's a fascinating possibility.

The Anti-Whimper

How many poets can you name? In my case, the answer is regrettably a rather short list. Perhaps your list is longer. But however long or short a list of poets anyone draws up, the name of T. S. Eliot is probably on it somewhere near the top. And his 1925 "The Hollow

Men," of which "Not with a bang but a whimper" is the last line, is one of his most famous poems. The penultimate line is "This is the way the world ends." But how did the (living) world begin?

As we saw at the start of this chapter, it didn't begin with a bang. There was no equivalent, for terrestrial life, of the hot big bang through which the universe is thought to have arisen. Life here began with a long wobbly trek from small molecules to big ones to collections of big molecules to protocells to cells, doubtless with lots of lineage extinctions along the way. Even by the time the first true cell appeared, after perhaps 100 million years of pre-cellular evolution, these primitive life-forms lacked many of the characteristics that we associate with animal life today, including, for example, the ability to see and hear, and the ability to produce light (bioluminescence) and sound (for example, bird song). These ancient organisms were soundless, like a snowdrop, not vocal, like a rooster. So why do I think of them as making a noise? Of course, the opposite of a whimper could just be silence. But that's not what I was thinking of when I gave this chapter its title. Rather, I was imagining a sort of noise that conveyed the opposite of what the whimper of a dying dog conveys: the beginning of life as opposed to the imminent end of life.

Suppose for a moment that those ancient microbes from which we all stemmed had been able to make noises; further, suppose that they magically knew about what some of their great-great-grandchildren would become billions of years later. What noises would they have made? Maybe a shriek of delight. Or, if able to articulate, maybe a brief exclamation like "Here we go!" Or even, to quote Buzz Lightyear, "To infinity and beyond!" In any event, it would be a noise that would reverberate far into the future, though perhaps changing in tone over time in the same way that early cosmic radiation changed over time into microwaves.

All of "us" (in the sense of all terrestrial creatures) are connected up in the great tree of life that has been ramifying ever outward since LUCA. We are all children of that same parent. Our lineages started off together and only later went their separate ways. But replace LUCA with a zygote and we have a similar arrangement of branching lineages, this time of cells rather than organisms. This cellular branching system can have different outcomes depending on where it's nested within the overall evolutionary tree. In one small corner of that tree it produced both you and me, and indeed all of our human brothers and sisters.

Chapter Nine

Our Internal Evolution

The Evolution of a Word

Species evolve; so too do languages. And for students of linguistics this is a very interesting process. For our purposes here we can ignore most of linguistic evolution, but we should make an exception for a single word, which is, fittingly enough, *evolution*. Today its meaning is clear and unambiguous, at least within the context of biology. The same was true in Charles Darwin's day—the nineteenth century. However, strangely, the two clear meanings are different. In Darwin's day *evolution* meant embryogenesis, or development, while *descent with modification* referred to what we now call evolution.

Evolution and development are processes that are both similar and different: similar in that they produce diversity of species or cells, different in that one is driven by natural selection and the other is not. Also, they are interrelated because, when a significant evolutionary change takes place, the whole life cycle including embryogenesis is altered, not just the adult. And often, though not always, the early stages of embryogenesis evolve more slowly than the later ones, with the result that in a comparison of, say, the early embryo of a human with that of a chimp or other mammal, the differences are very slight compared to those that become visible later on. Here's a quote emphasizing this point from the famous nineteenth-century German biologist Ernst Haeckel (when reading it, remember that *evolution* meant embryogenesis then): "The fact is that an examination of the human embryo in the third or fourth week of its evolution shows it to be altogether different from the fully developed Man, and that it exactly corresponds to the undeveloped embryo-form presented by the Ape, the Dog, the Rabbit, and other Mammals" (1896, 18). Apart from Haeckel's "exactly," this statement is true.

The shifting meaning of the word *evolution* has probably had an influence on the way in which the relationship between the two processes has been thought about. In fact, there's a book specifically devoted to this—*The Meaning of Evolution,* by Robert Richards. Here, though, we'll focus on a particular aspect of the similarity between evolution and development: the fact that both can be represented by treelike diagrams. And we'll think about our own developmental tree in a way that's ultimately inspired by Darwin, whose *Origin of Species* contained only a single diagram among its many pages—one that was an abstract evolutionary tree.

Things Cells Do

Embryogenesis is, among other things, a tree of cells. What I mean by this is that if we start with the single cell that we call the fertilized egg or zygote and think our way forward into the creature that ultimately develops from that zygote, we can picture the process as one in which cells keep dividing and producing daughter cells in an irregular, treelike pattern. It's irregular because, at any one time, some cells divide more than others—indeed, some cells die without producing any offspring at all. But overall it's treelike because new cells are always produced by divisions of old ones, never by the opposite of this process, which would be fusions of old ones. Therefore, throughout the process, cell number keeps going up, all the way from zygote to adult. Then it's brought back to a single cell again at the start of the next life cycle. Some animals—such as parasitic flukes—have more complex life cycles in which decreases in cell number happen more than once. But humans and other mammals have what are often called simple life cycles—ironic because, as adults, we're the most complex of all animals.

But cells do other things as well as divide, and we should take a brief look at these before returning to the cellular tree of embryogenesis. All cells have the job of staying alive long enough to do whatever their job is. Brain cells think, muscle cells contract, and skin cells provide a barrier between us and our external environment. Or, to be a bit more precise, these three types of cells collectively form structures or organs that do these three things.

In addition to dividing and staying alive, some cells can also *move* and *differentiate*. An obvious example of cell movement is the bloodstream. On a moment-by-moment basis, both red and white blood cells are moving through our arteries and veins. If they stop doing so, we die. But the cells of the blood vessel walls do not move. They stay in the same place, attached to the same neighbor cells, until they die and are replaced with other, equally static cells. This combination of mobile and static cells is essential for our existence.

Cell movement is important in embryogenesis too. At the early embryonic stage called gastrulation, which we looked at briefly in Chapter 3, lots of cells flow into the interior of the embryo. These produce, among other things, the gut. If they didn't flow inward, our gut might be on the outside rather than the inside, though no doubt any mutation that had such a drastic effect would be lethal to the embryo and would arrest its developmental progress long before birth. So we never see any humans with external guts, just as we never see any humans with brains in their feet.

What about that complex but crucial process that we call cell differentiation? In early stages of development, such as the mulberry stage, the cells are all rather similar; there are not yet any nerve cells, blood cells, or muscle cells. These specialist cell types are produced later. We noted in Chapter 3 that intercellular signals are involved in the development of the nervous system; they are likewise involved in all other forms of cell differentiation, whatever the nature of the cells that result. The process of differentiating involves some genes being switched on and others being switched off. For example, muscles have lots of the proteins actin and myosin, which enable contraction; brain cells don't. This is due to the switching on (or off) of the genes that make these proteins.

We saw earlier that the human genome contains about 23,000 genes, and that a complete copy of this genome is contained within every cell in our bodies, with only a very few exceptions (such as red blood cells). So in that respect all cells are genetically equivalent. But now we see that only a subset of genes are actually switched on and making proteins in any particular cell type. In that sense cells are genetically different.

Let's make a brief digression here to think about Dolly the sheep—the first cloned mammal—and also Freddy the frog, who should be more famous than Dolly, but isn't. Freddy is the name I've decided to give to the first *animal* that people succeeded in cloning—a title much more inclusive than the first such *mammal*. By "people" I mean a research group at Oxford led by the British biologist Sir John Gurdon. A horribly simplified version of what Gurdon's group did to make Freddy, and what Ian Wilmut's Edinburgh group did to make Dolly, is as follows.

You take an egg (from a sheep or a frog) and you kill off its nucleus, thus getting rid of all the gene-carrying chromosomes. Then you take another cell—this time a differentiated cell from the later development of the same type of animal, such as a sheep or frog skin cell. But now you keep the nucleus and get rid of the rest of the cell—that is, the jelly-like cytoplasm that is most of the cell's bulk and fills the whole space between the central nucleus and the outer cell envelope, or membrane. Now you make a hybrid cell by putting together the egg's cytoplasm and the differentiated cell's nucleus.

What happens is truly amazing. You get a tadpole (and eventually a frog) in one case, and you get a lamb (and eventually a sheep) in the other. Agents in the egg's cytoplasm kick-start development in both cases. These agents are able to start a process through which genes that have been switched off in the differentiated cell's nucleus are switched back on again—thus proving that they are still there and that cell differentiation only involves their (reversible) inactivation rather than their (irreversible) destruction. Gurdon was awarded a Nobel Prize for producing Freddy. Strangely, it took the Nobel authorities even longer than usual in this case—the work was done in the late 1950s and early 1960s, and the prize was only awarded in 2012.

You as a Tree

In biology, a yew is considered to be a tree but a human is usually not. Nevertheless, the "you" in my heading is not misspelled. Indeed, it could equally have been "me," or even "us," though in the latter case "tree" would have to be "trees."

The tree I'm referring to here was mentioned a little earlier on: it's the treelike pattern of cells giving birth to other cells that underlies embryogenesis. All the many trillions of cells that now make up *you* were produced as descendants of the fertilized egg cell that was your very beginning. Of course, it's a long road from zygote to adult, and the number of cell generations is too great for our cells of today to be called the great-great- . . . -great-grandchildren of our zygotes without having an astronomical number of greats. But that doesn't stop the mutual descent of *all* our cells, and the parent-offspring relationship between each of them and others, from being true.

Evolution can be looked at backward. For example, if we take modern *Homo sapiens* and trace its lineage back through time to our last common ancestor with chimps, as we did in an earlier chapter, we recover a single lineage and ignore the rest. Development can be looked at backward too, with the same result: we recover the ancestry, or lineage, of an individual adult cell. And the word *lineage* is indeed used by developmental biologists, thus reinforcing this parallel between the hierarchical processes of evolution and development.

Take your red blood cells, for example. We've seen that these are a bit odd because they lack a nucleus. But if you trace the lineage of one of these cells backward, you'll find that the loss of a nucleus occurs only at its end. Red blood cells are made in the bone marrow by *stem cells,* meaning cells that have the potential to produce offspring cells of varied types. A red blood cell is clearly not a stem cell, because, having lost its genes, it can't make any descendant cells at all. But white blood cells, which retain their genes, are not stem cells either. Rather, both of these are the terminal products of stem cells that have less familiar names, such as hemocytoblasts. As in most cases of technical jargon, you can chop up the name to see what it means. Here, *hemo-* means "blood" (as it does in the blood protein hemoglobin), *-cyto-* means "cell" (as it does in *cytology,* another name for cell biology), and *-blast* refers to development (as it does in the case of the developmental stage that comes immediately after the mulberry stage, the blastula). In fact, *-blast* comes from the Greek *blastos,* "sprout," and is thus part of the name of many types of stem cell.

Now let's go back even further in the life of a red blood cell. Its immediate ancestor was only capable of making one type of cell; the ancestor of that one was capable of making more than one cell type, but still only types of *blood* cell. As we go back further, we encounter stem cells that can make an even greater number of different types of descendant cell. And when we get back to the early embryo we have reached cells that are capable of making offspring cells of any type at all—these are referred to as totipotent embryonic stem cells.

To read the whole cell-lineage tree backward we would have to take many different starting points. After all, starting with an adult red blood cell and going back to its distant ancestors in the embryo only answers the question of where that particular cell type comes from. What about the others? Well, the journeys that we'd embark on if we traced nerve cells or muscle cells back to the early embryo would be different in detail but similar in essence. And since this book is about essences rather than details, we won't trace all those variant journeys back to the egg. But we should briefly consider how many cell types there are in a human (and in other animals), for otherwise we don't know how big a fraction of a body we traced back when we took the red blood cell's route.

A Multitude of Fates

As development proceeds and cells differentiate in various ways, how many possible fates are there? In other words, how many cell types are there in an *adult?* The ballpark figure usually given in answer to this question, when asked about a human or any other mammal, is about 200. But you could say that the question is a bit like the proverbial "How long is a piece of string?" It's possible to divide cells up into categories, such as blood, brain, and skin cells, then to divide each of those further, for example into red and white blood cells, and then to subdivide into still narrower categories—though we might run into a brick wall in some cases. For example, "red blood cell" is really just a single type, while "white blood cell" is a category containing several types.

The "about 200" answer for a mammal refers to the number of cell types that we end up with when we've narrowed each category down to a single type—though bear in mind that there will always be a biologist somewhere who will argue that "type x" is really a mixture of hitherto unrecognized subtypes x1 and x2. The number of cell types in an adult animal that is not a mammal depends very much on the group to which it belongs. Birds have a similar number of cell types as mammals. But many invertebrates have only about 50. And the simplest extant animals, sponges, have only about 10.

The production of different types of cell is a really important part of our developmental process. You couldn't be a human with a mere 10 cell types; it's simply not possible. Nevertheless, differentiation is only a *part* of development. There's what might be called a complementary part, the part that ensures that the brain ends up in the head and not in a foot. But since the foot-brained human is only a theoretical possibility that sounds rather ridiculous, let's approach this other aspect of development from a different angle.

Fingers and Muscles

If you haven't already read it (or seen the film), I'd strongly recommend *My Left Foot,* by the twentieth-century Irish writer Christy Brown, who grew up in a poor, and very large, Dublin family. He was afflicted by cerebral palsy and was unable to use his hands. The book was written, as its name declares, with the toes of his left foot. This is a classic story of triumph over disability. Writing a book is hard enough work when you can use your hands. Having to make do with one foot makes the job almost impossible, but Brown did it—and he went on to write more books, and to paint pictures too.

The difference between hands and feet partly resides in the difference between fingers and toes. From the perspective of the cell types involved, they're all the same. Also, the thumb is the same as the fingers from this point of view; and the fingers—index, middle, ring, and pinkie—are the same as each other. But from other perspectives they're not. Some people have bigger differences between their fingers than others. Often, the smallest difference, looking at the fingers "from the outside," is between the middle and the ring. Yet even these are not identical. And the difference between either of them and the thumb is very pronounced indeed, to the point where the thumb has one less joint.

Like everything else in development, the making of fingers sometimes goes wrong. Some people have too many fingers (polydactyly), others the wrong sort of fingers—for example, the thumb can end up with an extra joint. And some unfortunate folk have both problems together—the wrong number and the wrong type. For example, the index finger can be a second thumb and the other fingers can be missing altogether. I saw a person with this condition at a fair in the United States when I made my first trip there as a student on a working vacation. Outside the booth where this man was, I noticed a sign with a picture of a person with crab pincers instead of hands, and I assumed it was some sort of stunt—a bit like Edward Scissorhands. But when I paid the entrance fee and went in, I saw a person whose hands terminated in two opposable thumbs and no other digits. I came out with very mixed feelings; I didn't know whether to regard the whole thing as a case of appalling exploitation on the part of the fair owners or as incredible entrepreneurship on the part of the "crab-man." I would have preferred it if he had written a book, like Christy Brown; but then again, not everyone's a potential writer.

The message of this sad story is that when things go wrong in development they teach us something about how development works. They point to the existence of mechanisms that normally produce the right result; if we never saw the wrong result, we might not pause to consider these mechanisms.

We can arrive at the same point by starting with muscles rather than fingers. Think about Popeye, the famous cartoon sailor who had very pronounced biceps muscles, especially after eating spinach. Although the rest of us have slimmer biceps, they're of the same general shape—it's called a spindle shape. But we have many other muscles that are very different shapes. For example, we have muscles in our foreheads that enable us to raise our eyebrows. These muscles are flat; if they were spindle-shaped, we'd look a bit odd.

So the key question is this: for every sort of large structure that occurs multiple times in the human body, with the different instances of it being different in size, shape, and other features despite being made up of the same types of cell, how are these differences reliably produced, with only very occasional exceptions? We're no longer talking about making the right cells, but rather about organizing them correctly in space. This puzzle is not fully solved yet, but we're making headway.

Another version of the same problem is the making of two mirror-image copies of the same thing. For example, your right thumb looks quite similar to your left, but if you think about its connections to the rest of the hand, you'll recognize that they're reversed in one case compared to the other. And your kidneys are mirror images of each other. They're a bit like a pair of fat commas facing inward, but if things go wrong they could face outward, or take some other shape entirely. So again, how does the normal arrangement almost always result?

Developmental biologists put this general type of puzzle under either or both of the headings *pattern formation* and *morphogenesis*. Literally if not biologically, these are synonymous. After all, a shape is a pattern in 3-D space. And the generation of a form (Greek *morphos,* "shape") is the same as its formation, as the words themselves show. But putting a name to a process isn't the same as giving an explanation of how it works; naming is only a start.

Genes are key players, as you might expect. Quite a few genes are known that control various aspects of pattern formation; if these mutate, the wrong pattern is generated. Over the last couple of decades quite a lot has been learned about these genes in various creatures. We now know about genes that control the left-right orientation of various aspects of human development. The medical condition called *situs inversus,* wherein multiple internal organs are transposed from left to right (but the person looks completely normal from the outside), can be caused by a defect in one of them. A similar gene in snails can reverse the direction of coiling of the shell from right-handed to left-handed. Indeed, finding more or less the same gene doing more or less the same thing in very disparate animals has become commonplace now, though when it began to happen it was one of the most exciting discoveries of "evo-devo," or evolutionary developmental biology. For a good account of this, see *Endless Forms Most Beautiful: The New Science of Evo-Devo,* by American biologist Sean Carroll.

Inside the Abdomen

There was a strange habit among early biologists of naming parts of the bodies of very different animals in a similar way even though they are not *homologous,* that is, evolutionarily descended from the same structure in an ancestor. For example, it makes sense to call the two long bones of the human forearm and their equivalents in mice the radius and ulna, because we both inherited these, in modified form, from the last common ancestor of our two species. But to call a human leg bone a tibia and then find that part of a fly's leg is also called a tibia, despite the mammalian and insect legs being constructed on entirely different bases, is a bit weird; yet it happens. And what applies to the legs applies also to the trunks they're attached to.

Think of a large queen wasp. She has a broad head, a thin neck, a broad thorax, a thin waist, and a broad (and long) abdomen. It's her abdomen that is striped with warning coloration, probably because it's the biggest part of the body and thus the most conspicuous—hence it's the best way of signaling danger to would-be predators. Although most of us are familiar with a wasp's abdomen from the outside, few of us have seen it from within. It's too small to dissect easily in recently killed specimens. So, although it contains important wasp organs, we don't have a clear mental picture of them.

In contrast, the organs within a human abdomen are well known to most of us, usually from diagrams in books or online. Almost all high school biology textbooks contain such pictures. And in my case I have mental images of the inside of the human abdomen not just from pictures but from the real thing, because I started out as a student of medicine before deciding to switch to science. This means that I got the opportunity of assisting with the dissection of a body belonging to one of those noble people who donate theirs to medical education. My first view of the layout of those tightly packed organs—stomach, liver, spleen, pancreas, intestines—is something I'll never forget.

The overall picture—seeing organs—is the same in the thorax (heart, lungs) and the head (brain) as in the abdomen. It is the organs, not the cells, which are the obviously visible units that make up the body. We know that organs are composed of cells, but it's the organs themselves that strike the observer, whether surgeon, medical student, or textbook reader. They constitute the top level of the makeup of our bodies.

Now, this might seem like a bizarre question, but here goes anyhow: what are the equivalent top-level entities in the makeup of the universe? If we think of the whole of the observable universe as a single entity, like a body, what are its "organs'? Of course, the universe is not a vast organism (as far as we know), so the parallel shouldn't be overinterpreted, but it's interesting nevertheless.

Will the answer to the above questions take us full circle to where the book began—with galaxy gazing? The short answer is yes and no; now let's look at the longer one.

STRUCTURES AND FUNCTIONS

Chapter Ten

Spacious Heavens

Flies in Orchards

We're now going to look at the highest level of structure in the known universe. And a good starting point, though not the most obvious one, is the spatial distribution of flies in orchards. Or fish in the sea. This is because, whether we're dealing with galaxies or animals, the focus of interest in this chapter is how things are arranged in space. There are some key ideas on this matter that can be applied to any objects distributed through the three dimensions of space, regardless of the nature of the objects or the spatial scale involved—small, middling, large, or beyond. It makes sense to begin our exploration of these ideas in relation to objects that are familiar, like flies or fish, in a realm of space that's also familiar, the fairly small, before tackling unvisited objects in the realm of the utterly vast.

Imagine a small island that's entirely given over to agricultural land: mostly fields but a few orchards too. Animals of many sorts inhabit the island, from the worms in the ground to the birds in the trees. Let's take just one animal species out of the lot—a fruit fly that lays its eggs in rotting apples, which serve as a food supply for its larvae—and ask the question: what would a picture of its spatial distribution across the island look like?

Let's consider this picture as taking the form of a large map of the island with individual flies being represented by dots. The pattern of dots would almost certainly take the form of a series of clusters, because there would be more flies in the apple orchards and fewer in the fields of crops. And not all orchards would be equally populated—whichever of them has the most rotting fruit would have the densest population of fruit flies. So if we could magically see such a map—imagine the practical difficulties of producing it—we'd see what ecologists call a clumped distribution. Astronomers tend to talk about clusters rather than clumps, so we'll use that term since we're headed for galaxies. But let's not leave our island just yet.

Within the "best" orchard, and thus within the densest cluster of flies, there will be clusters on a smaller scale. Flies laying eggs will be clustered on particular rotting apples that have fallen to the ground. So there are clusters within clusters. This type of pattern is common in the animal world. It's found not just on small islands but on larger landmasses and indeed in the seas that surround them in the case of marine animals such as fish.

Imagine now a hologram of a volume of sea with all the fish of a particular species shown as dots against a 3-D blue background. Some parts of the sea are likely to be more favored than others, perhaps for reasons to do with food supply. These are where the most fish will be found; so again, as with the distribution of flies, we can see clusters, but now in three dimensions rather than two. And within each of these high-density parts of the sea, each representing a cluster of fish, there may be shoals of the type of fish concerned, with relatively few fish swimming singly or in pairs a long distance from the shoals. A shoal is a cluster, so again we have the clusters-within-clusters pattern. This pattern is not universal in the animal world, but it's overwhelmingly common.

Going Up

Now we have to make quite a mental transition—from thinking about animals to thinking about galaxies. Let's make the transition easier by doing it in stages: planets are a good intermediate. Consider the differences in scale between a whale, a planet, and a galaxy. These three objects fall into size categories that we can call, respectively: large, enormous, and vast. But is the difference between large and enormous here bigger or smaller than the difference between enormous and vast? If we do the sums to work out how many whales make a planet-sized object and how many planets make a galaxy-sized object, we find that planets are much closer in size to whales than they are to galaxies.

So how are planets distributed? Well, as we all know, there's a cluster of eight of them around our Sun. It's necessary to go a very long distance through space before we get to a second cluster surrounding another sun. And then another very long distance before we get to a third. But are planets simpler than animals in their distribution? Is there just one level of clustering? Or does the clusters-within-clusters pattern apply to planets too?

Let's bring poor old Pluto to our aid here. This is the rocky body in the outer reaches of our solar system that was previously a "real" planet and was then reclassified as just a dwarf planet, ironically in between the 2004 launch of NASA's *New Horizons* probe to Pluto and its arrival there in 2015. Pluto was reclassified because it fails to meet the criterion of planetness that says a planet must be alone in its orbital zone around the Sun. Not alone in the sense of lacking moons, of course, but alone in the sense that there are no other large bodies near it that are orbiting the Sun rather than orbiting the body in question—in this case Pluto.

There's another criterion of planetness that Pluto *does* satisfy: being approximately spherical. This is not enough on its own for us to call something a planet; if it were, then the Moon would be a planet, whereas instead, because it orbits the Earth, it's called a satellite. Interestingly, although many other planets, for example Jupiter and Saturn, have spherical moons, the two little moons of Mars, Phobos and Deimos, are irregular in shape and are not even close to being spheres.

The key word in the last sentence is *little*. This is another case where size matters. A body in space that's above a certain mass becomes approximately spherical because of the effect of its own gravity. You can think of this as a strong gravitational attraction pulling every part of the surface inward by the same degree. Below the mass threshold, however, there isn't enough gravity to do this, so irregular shapes are common for small rocky bodies in the solar system—for example, most of the asteroids, as well as the small moons. Mars's two moons are well below the threshold, each being less than 1 percent of our Moon in their average diameter.

Let's for a moment use sphericalness as our *only* criterion for bodies in the solar system to be "planets." In this case it becomes clear that the solar system does indeed have clusters. The Earth, for example, is a cluster of two. Jupiter and Saturn are larger clusters because they each have multiple spherical moons. This approach leads us to realize that our solar system *does* have clusters of bodies; it's just that the standard definition of a planet prohibits clustering of planets in the sense used temporarily here. If we talk instead of spherical bodies, there most certainly is clustering; and since each solar system is itself a cluster, we're back to that old animal pattern of clusters within clusters.

And So to Galaxies

We now make the second mental transition—from planets to galaxies. How are these distributed in space? Science has come a long way toward answering this question over the last century. In 1920, what's called the "Great Debate" took place between the American astronomers Heber Curtis and Harlow Shapley at the U.S. National Museum (now the Smithsonian Museum of Natural History) in Washington. Curtis believed that there were other galaxies beyond our own Milky Way, whereas Shapley believed that the claimed "other galaxies" were just objects within the Milky Way—a view which, if correct, would mean that the Milky Way was in fact the entire universe. It turned out that Curtis was right, Shapley wrong, and we now know that the Milky Way is just one of many billions of galaxies. We owe the resolution of the debate to another American astronomer, Edwin Hubble, who used what was then (in the 1920s) a novel method to measure the distance to Andromeda, thereby proving that it was well beyond the outermost fringes of the Milky Way.

In fact, Andromeda is the closest large galaxy to our own. And we already put a figure to its distance in Chapter 1—about 2.5 million light-years. Very close indeed, at least in intergalactic terms. The so-called local group of galaxies, of which the Milky Way and Andromeda are constituents, consists of a cluster of about 50 galaxies, covering a distance of about 10 million light-years. Most of these, and indeed possibly most galaxies in general, are "dwarf galaxies," meaning that they contain only a few millions or billions of stars as opposed to the hundreds of billions that characterize a full-size galaxy like our own.

So, galaxies exist in clusters, at least in our own locality. But what about elsewhere? And do clusters exist at two or more levels the way they do for animals and planets? The answer to the first of these questions can be found in what has become referred to as the Copernican principle. This is the idea that our own location in the universe is not special. In the history of human thought, both about the living world and about the heavens, we've taken a long time to shake off our arrogant philosophical starting point that somehow we're at the center of things. At last we're free of that misleading worldview—well, some of us, anyhow.

If we believe the Copernican principle, there should be groups, or clusters, of galaxies all over the place. And, now that we can see far enough into the universe, we know this is indeed true. Let's turn to the remaining question, then: are there clusters within clusters?

Indeed there are. Our local group is part of a larger collection of galaxies called the Virgo cluster. Here, instead of there being a mere 50 or so galaxies, as in our local group, there are more than 1,000 of them. So we find a pattern of clusters within clusters at the level of galaxies, just as we found this pattern at the levels of animals and planets. But is there another level of cluster beyond the two that we've already seen? Yes indeed, and it's called, unsurprisingly, the level of superclusters.

Laniakea

Our local cluster has been named the Virgo cluster for many years. Our local supercluster used to have the same name, which was rather confusing. But as of 2014, our supercluster has both a new, more inclusive boundary and a new name—Laniakea, which means "spacious heavens" in Hawaiian. This is an excellent choice of name: the distance it spans is about 500 million light-years, and the number of galaxies within it, though hard to estimate, is somewhere in the region of 100,000.

The shape of Laniakea is usually pictured as something like a hairy twig; the general form of this twig is shown in the illustration at the start of this chapter (page 112), with a cosmic butterfly climbing on it. But how are we to interpret shape at such an enormous level of spatial scale, a scale that's far beyond anything encountered in everyday life?

We got to galaxies from a starting point of animals. Likewise, we can use familiar things to get to galaxy superclusters, but this time those things are rivers. In northern England, where the country is divided into eastern and western portions by the "backbone" of the

Pennine Hills, there are western drainage basins that run into the Irish Sea and eastern ones that discharge into the North Sea. In each case, multiple rivers converge as tributaries of another, and the whole river system discharges into the sea via its estuary. For example, there's a drainage basin in northeast England that discharges into the North Sea via the river Tyne. And there's one in northwest England that runs into the Irish Sea via the river Mersey.

How do river systems help us to understand galaxy superclusters? Well, the connection is movement, or flow. On the surface of the Earth, movement is a relatively simple concept. The river Tyne flows but its banks do not. They stay very still from our human perspective, even though we know that, from an astronomical perspective, they're rotating very rapidly, along with every other topographical feature on the Earth's surface. We detect movement of the river in relation to the apparent stasis of the ground on either side.

Movement is a trickier concept in space. The Earth orbits the Sun. But does the Sun orbit anything, and if so what? And at higher levels, does our galaxy, or the cluster or supercluster to which it belongs, move? If so, relative to what?

Clearly these are difficult questions, but they're not unanswerable. The universe in general is expanding, as we've already discussed. On a large scale, the pattern of expansion is uniform. It's called the Hubble flow. But when the movements of individual galaxies or groups of galaxies are studied, we find that these cannot be accounted for solely by this flow. Rather, galaxies and groups of them move partly with the flow and partly relative to it. The latter type of movement is called *peculiar.* And it's the peculiar movements of galaxies that are used to define our supercluster and to delimit it from neighboring superclusters.

Within Laniakea, each galaxy or local group of galaxies has peculiar motion in a particular direction. Typically, this direction is toward the center of mass of the supercluster—the so-called Great Attractor. Beyond the periphery of Laniakea, galaxy groups flow in other directions. This is the basis for recognizing our supercluster—and indeed others.

The universe consists of many superclusters. This raises the question of whether an even higher level of clustering exists—super-superclusters, if you like. At the moment, it seems not. Instead, when we look for structure above the level of the supercluster we find relatively one-dimensional thread-like *filaments,* two-dimensional *sheets,* and empty areas called *voids.* Thus at the highest level of all, the universe resembles a gigantic spiderweb. It probably wasn't made by a giant spider, but I could perhaps argue that such a hypothesis of its origin is no worse than the alternative hypothesis of the big bang as a causeless event.

Any brief account of such colossal matters as the one in this chapter is necessarily a simplification. There are all sorts of complexities, but I don't think they interfere with the overall picture. For example, despite the existence of groups and clusters of galaxies, some galaxies seem to exist more or less on their own. These are called field galaxies; each of these is a bit like an individual fish that is on its own, midway between two shoals. Reality is messy at pretty much every level. This shouldn't stop us from searching for general patterns, as long as we're prepared to acknowledge the possibility of exceptions.

The Fourth Dimension

Pages and screens are two-dimensional, so all illustrations of Laniakea that you can find—for example, using Google Images—are necessarily simplified views. In fact, they are 2-D slices through a 3-D structure. Moreover, most or all structures that we can think about are in fact 4-D—the only exceptions would be structures that do not change at all over time, and it can be argued that these don't exist. Our bodies are four-dimensional structures. So are the houses we live in. Houses tend to have longer life spans than people, but not always. The house I grew up in was bulldozed by the people who bought it after my parents died. Its life span was 62 years, whereas my mother, who lived in it for most of her life, reached the grand old age of 89. Planets have life spans too. The Earth is about 4.5 billion years old and will be either obliterated or drastically altered about the same amount of time into the future, when the Sun becomes a red giant. So, what about galaxies and their clusters?

This question deals with a subject that is perilously close to the frontier between knowledge and ignorance—or between our islet of understanding and the vast ocean of the unknown to which T. H. Huxley referred in the quotation given at the start of the book. Much present-day research is being conducted to try to answer it. Popular science books have to tread especially carefully when dealing with such topics. The best strategy is to be brief but to try to convey the essence of what we're currently grappling with. That's the aim of the next two paragraphs. So take this account of the origins of galaxies with a grain of salt; the story will change as we learn more.

We think that the large-scale structure of the present-day universe arose from a combination of small-scale structure in the early universe and its subsequent expansion. During this expansion, slight clumpiness in the distribution of matter became enhanced by the effects of gravity. However, an important factor in this transition from small-scale to large-scale structure may have been the coming into existence of dark matter at an early stage in the universe's life. This enigmatic stuff is not yet understood, but it is thought to consist of a type of fundamental particle that we still haven't discovered, rather than being composed of protons, neutrons, and electrons, like "normal" matter. At the time of writing, these dark matter particles are referred to as WIMPs—weakly interacting massive particles—and they remain to be observed anywhere, whether on Earth or in any of the close, middling, or far realms of space. One day they may emerge from the realm of the hypothetical to the realm of the actual, but they haven't made this transition yet.

Despite that fact, it's possible that the distribution of dark matter played a gravitational role in determining the origins of the structure of ordinary matter. But that's as far as I'm going to go in terms of galactic origins. The question has now become: does the distribution of the matter we can't see predate, and somehow *cause,* the distribution of the matter we can see? A lot of recent evidence suggests that this is indeed the case. But this issue is still far from being resolved. So we'll leave the question hanging and turn to a final one before we proceed to Chapter 11.

Structures without Functions?

Let's return momentarily to the familiar world of Earth-bound creatures. In biology, we often talk about structure and function. The branches of biology that deal with these are anatomy and physiology, respectively. Any organ that we choose to study has a structure and at least one function. For example, the heart is, well, heart-shaped, and pumps blood. The biceps muscle is spindle-shaped and, along with other muscles, causes the arm to be able to lift things by flexing the elbow. The stomach is a roundish bag with the job of carrying out some of the digestion of our food. We don't expect to find a major structure in our bodies that has no function at all. Minor structures are different; think of a vestigial structure, such as the appendix, which is probably evolutionarily on the way out.

Deep down within any organ, the cell types of which it is composed all have particular structures and functions. Within a muscle, for example, there are various types of blood cells as well as muscle cells. These can easily be recognized because their structures are different from those of the cells of the surrounding muscle tissue, and these different structures reflect their different functions. Again, as with organs, we don't expect to find a type of cell in the body that has no function whatsoever.

If we shift our attention from one group of organisms to another, the story stays the same. In the preceding paragraph all my examples were from the animal kingdom. But the situation in the plant kingdom is similar. Now let's make a bigger scientific switch than the one from zoology to botany. Let's move from biology to geology. Rocks have structures too. And, like the structures of organs, these can vary from one case to another. Some cases are particularly striking. For example, on the north coast of Ireland and the west coast of Scotland there are massive hexagonal columns of basalt, some of them several meters tall. The areas where they are found are called the Giant's Causeway (County Antrim, Ireland) and Fingal's Cave (island of Staffa, Scotland). These are very impressive, and very particular, structures. But do they have a function? Indeed, is that question even a meaningful one?

The answer to both the above questions would appear to be no. And the answer to whether, at a much larger scale, galaxy clusters have a function is probably no too. But contemplating such questions is still useful because it forces us to think about what we mean by a function. Parts of animals have functions because they help the animals to survive. Parts of plants have functions of the same survival-related sort. But "function" is not solely the preserve of living things. The keyboard that I'm using to type this has a function. So does the resultant book. Neither of these is alive. So what is it that they have in common with life-forms that they don't have in common with rocks? This question is perhaps leading us into a subjective realm, so please consider what follows as just my personal view; scrutinize it from every angle to see if you agree with it.

If something can be said to have a function, then it's something that has been made by a life-form. It needn't be alive itself; my keyboard is an example. And it can be alive but not conscious; a flower, for example. The human brain is an example of a structure that is alive, conscious, *and* made by a life-form—a human embryo, which was in turn made by its parents. But it's a very special case. Most structures that have functions are not conscious.

Does a structure that has a function always have something to do with survival? Probably not. Although spacesuits and submarines have a lot to do with the survival of the people who inhabit them, I would survive all right without my keyboard. Human-made objects with functions can probably be divided into those whose function is for survival, progress, moneymaking, amusement / enjoyment, or some combination thereof. The keyboard can be used for three out of the four. For most people, a bar of chocolate can only be used for one of these; but for the owners of Cadbury's two, and for someone starving a different two.

Arguments about function were used extensively in the early 1800s for theological reasons. Archdeacon William Paley was a churchman who had a parish in the north of England, at Bishopwearmouth, Sunderland, and was also subdean of Lincoln Cathedral. The opening sentences of his book *Natural Theology,* published in 1802, have become famous and are well worth quoting here:

> In crossing a heath, suppose I pitched my foot against a stone and were asked how the stone came to be there, I might possibly answer that for anything I knew to the contrary it had lain there forever; nor would it, perhaps, be very easy to show the absurdity of this answer. But suppose I found a *watch* upon the ground, and it should be inquired how the watch happened to be in that place, I should hardly think of the answer which I had given, that for anything I knew the watch might have always been there. Yet why should not this answer serve for the watch as well as for the stone; why is it not admissible in that second case as in the first?

Paley's answer, of course, was that a watch is a complex object that has a function, and thus must have been *made* by somebody or something. And to Paley humans were like watches rather than stones.

Paley's book had a big influence on Charles Darwin, though probably not the one that Paley would have liked. It also, much later, led the English biologist Richard Dawkins to come up with the memorable phrase and book title *The Blind Watchmaker*. As Dawkins explains, Paley's watchmaker is none other than Darwinian natural selection—operating blindly rather than in a goal-directed way. Paley seems to have inspired most the people he would have least liked to help develop their views.

Natural selection takes place in ecosystems. Each of these contains many interacting organisms. Organisms are made up of organs, which can all be said to have functions. But do ecosystems have functions? As we'll see in Chapter 11, this is a tough question.

Chapter Eleven

The Ecological Theater

Time within Time

The ecologist George Evelyn Hutchinson published a book in 1965 with the wonderful title of *The Ecological Theater and the Evolutionary Play.* This title beautifully encapsulates the relationship between ecology and evolution. All evolution takes place within ecosystems, wherever they may be on the planet. It's the stresses and strains of ecological interactions—with predators, competitors, climate—that give rise to Darwinian selection.

Ecosystem, the technical word for an ecological theater, simply means the totality of everything living and non-living within a particular area. In other words, all of the interacting organisms, together with all the other things that they have to interact with, like air, water, and soil. Or all of the actors and all of the scenery and stage. However, there isn't just one play going on in an ecosystem, there are two, and one of them takes much longer than the other, at least in terms of how they affect the players.

The distinction we're getting to here is that between ecological and evolutionary time. The former takes days, months, years, centuries. The latter takes millennia and beyond. There's no absolute marker between the two—a common situation in the biological sciences—but recognizing this dichotomy is helpful in terms of studying what happens out there in those natural theaters.

In ecological time, organisms interact with each other and with the environment in general, and as a result, their *numbers* change—they go up and down. But their *nature* remains essentially the same. In contrast, the longer-term play of evolution alters the nature of the players. As with the distinction between the two timescales, the distinction between the two processes isn't clear-cut. Sometimes evolution can happen unusually rapidly. This is especially the case in extreme conditions, and these are often produced by humans.

A classic example of such rapid evolution involved the insecticide DDT, which in the mid-twentieth century was indiscriminately sprayed across fields of crops. The insect pests that DDT was intended to control were indeed decimated by this toxic substance. But so too were other insects. DDT doesn't care if you're a crop-eating pest or a beautiful butterfly feeding on wild flowers at the field margins—it kills you equally well. Although the users of the insecticide may have had concerns about this issue, they had much more pressing concerns some years later when the DDT stopped working. The farmers had been unwittingly carrying out a field experiment on how Darwinian selection works, and they showed how amazingly effective it can be in causing the evolution of resistance to DDT, even on timescales that are only several generations long for the evolving insects.

Normally, though, evolution takes longer to alter the animals or plants concerned. For example, in the evolution of *Homo* in the ecosystems of sub-Saharan Africa a million years ago and more, those protohumans that eventually became *us* did not evolve significant differences within just a few generations. Their environments must have posed them many challenges, but none as severe as an environmental agent that could kill off more than 99 percent of them, as DDT was capable of doing in populations of insects.

For the moment, let's forget about evolution—we'll bring it back again shortly—and focus on the short-term goings-on within a particular ecosystem. We'll imagine a remote island of a few square kilometers, uninhabited by humans, which is completely covered in deciduous forest. This way, unlike on a continental landmass, there's no agonizing over the spatial extent of an ecosystem and the problem of having to decide where one system starts and another finishes. On a continuous tract of land this is impossible, except where there are abrupt habitat boundaries, for example between forest and field, and even here some animals may move between one system and the other. A small island provides the most abrupt boundary of all—that between a terrestrial ecosystem and its marine counterpart.

A Cartoon Ecosystem

Here's a commonly encountered picture of an ecosystem that both informs and *dis*informs. Trees and other plants capture sunlight and convert it into biological energy. Herbivorous animals graze on the plants—perhaps insects eating their foliage and birds eating their fruits or seeds. Carnivores eat the herbivores—for example, insectivorous birds eat the insects. And birds of prey, such as owls, eat the insectivorous birds. Eventually, all individuals belonging to all these categories die; their bodies are consumed by scavengers, such as rats, and microbial decomposers, thus turning what's in them into nutrients in the soil, which get taken in by plant roots, thus completing the ecological cycle.

All of the statements in the above paragraph are true. So why did I refer to it as a cartoon ecosystem, and why do I think that the picture thus painted includes *dis*information? It comes down to the difference between "the whole truth" and "nothing but the truth."

It's true that some plant material is eaten by herbivores—but how much? Think about walking through a deciduous wood in the summer and looking up at the beautiful foliage of the trees—oaks, sycamores, ashes. Are the leaves typically near perfect, or are most of them skeletal affairs from which almost all the fleshy tissue has been stripped by insects except around their veins? The former picture is closer to the truth. And it remains so if we consider the plants of the forest floor—grasses, nettles, brambles.

The opposite situation—leaves stripped nearly bare, for example by insects or slugs—can readily be found in a garden. But, much as we like gardens, they're not truly natural. The plants we grow for their beautiful flowers are often from other parts of the world and aren't well adapted to our climate. Also, many garden plants have been modified by breeders, who may have succeeded in making their flowers more brilliantly colored but may also have unwittingly modified other characteristics of the plant for the worse, in particular those related to survival.

So on our little wooded island, though maybe not in my garden, the norm is that most live plant material never gets eaten by herbivorous animals. Instead, most leaves survive more or less intact until autumn when they fall to the ground and begin to rot. The energy inside them goes not to flying herbivores like birds and dragonflies, but rather to their less photogenic counterparts in the detritivore system in and on the soil. It's more likely that worms and bacteria will get a leaf's energy than a caterpillar or a bird that eats caterpillars.

Of course, a tree is not just made of leaves. A lot of its bulk is made of wood and bark—trunks, branches, and twigs. But this material is rather inedible to most herbivorous animals, so its future—for example, that of a fallen branch—is also likely to lie in the hands, or mouths, of detritivores and decomposers. Some of a tree's energy is used in the making of reproductive structures such as pollen, seeds, and fruits. These are much more likely to be eaten by herbivores; indeed, it's in the tree's interests that they are. Plants actively use animals for pollination and dispersal. However, this stored energy is only a small minority of the total biological energy the tree has made by photosynthesizing.

What this means is that in the typical forest ecosystem about 90 percent of the energy garnered from sunlight by plants never enters the herbivore-carnivore chain at all. Such ecosystems are theaters in which the players are mostly plants, fungi, and microbes, along with certain detritivorous animals such as earthworms. As ever, we have to be careful about sweeping generalizations; blackbirds eat earthworms *and* herbivorous insects, so the energy pathway from a tree to a blackbird can take different routes. Thus it's clear that the separation into a herbivore-based system largely above the ground and a detritivore-based system largely beneath it is too simple. Nevertheless, many animals, and especially carnivores, are unnecessary extras from the ecosystem's point of view. We should not mistake the cartoon ecosystem of an aesthetically driven wall poster for its real counterparts in which most of the energy from plants bypasses carnivores altogether and ends up in the bellies of worms and other unglamorous dwellers of the interstices of the soil.

Thinking about Function

We know that talking about an ecosystem as having a "point of view" is only a figurative use of that term. In a more technical book on ecology such usage would be avoided. However, the term *ecosystem function* is widespread in the ecological literature. But what does this mean? Given our notion of function, in Chapter 10, as applying only in the realms of organisms and things they make (birds' nests, computer keyboards), can an ecosystem really be said to have a function at all?

Sometimes ecosystems are thought of as what could be called superorganisms. Taking this viewpoint to its extreme, the whole of the Earth and its atmosphere (including the totality of all earthly ecosystems) can even be thought of in this way—as in the Gaia hypothesis of British environmentalist James Lovelock. From such a perspective, a function of the ozone layer is to reduce the amount of harmful ultraviolet light reaching the Earth's surface, at or near which most plants and animals live.

The trouble with this way of looking at things is that it's too centered on *us* (in the sense of humans, animals, or life-forms generally). Our organs have evolved to their present form because they aided our survival. The ozone layer did not take its present form for any such reason. Equally, in a particular forest ecosystem, the microbes that cause decomposition can be said to have the "function" of releasing the nutrients that might otherwise remain locked up in fallen leaves or dead animals, and making them available to plants. But the microbes evolved on the basis of their own survival, not to perform a service for the greater good of the ecosystem as a whole.

The case of carnivores is especially interesting. I described them earlier as a sort of "extra" in an ecosystem. Another way of putting the same point is to say that the ecosystem would function perfectly well without them. The *routes* of energy flow would be a bit different, as would the patterns of nutrient cycling, but these processes would continue.

Let's focus on a particular type of carnivore—a parasitic wasp. These are little-known creatures outside the realm of biology. And yet there are thousands and thousands of them—many thousands of species, each with many thousands of individual wasps. They're usually very small, and they don't have conspicuous black and yellow stripes. So they don't look like what we might call ordinary wasps, yet they're closely related to these more familiar insects.

The best way to describe the life cycle of a parasitic wasp is to liken it to that of the human parasites that feature in the film *Alien*. The idea of a creature that develops inside us and then erupts out of our bulging skin, killing us in the process, is repugnant. It's lucky for us that such aliens don't exist—or if they do, then they're not (yet) here. But for some of our fellow creatures on this planet, the "aliens" are real, here, and now.

Although the exact details of the parasitic wasp's life cycle vary from species to species, here's the essence of what happens. A caterpillar, whose normal fate is to become a butterfly, is busy feeding on a leaf. A female parasitic wasp lands on or beside it and injects her eggs into it. The caterpillar's fate is now very different. It will carry on eating for several days, during which time the eggs inside it will hatch and the resultant larvae will begin to eat the caterpillar from within. As the wasp larvae grow, the caterpillar becomes progressively less healthy, though as long as it can it keeps on eating, thus unwittingly providing nutrition for the larvae that will soon kill it. One day, the caterpillar's skin begins to bulge; then it splits and out of it come one or more wasps. The caterpillar is dead, the wasps survive, and off they go to find other hosts.

Even more bizarre is the existence of insects that are called hyperparasites. These lay their eggs inside the larvae of a parasitic insect. Often, both species are types of wasps.

Now to the really interesting question: what is the *function,* within the ecosystem, of a parasitic wasp, be it an ordinary parasite or a hyperparasite? In my view, none whatsoever. These creatures are the ultimate extras. The ecosystem concerned would work very well without them. Again, the exact pathways of energy flow would be altered, but the ecosystem would get along fine. Like all other creatures, their function is to survive and reproduce. That's it. Any effects on the rest of the system, such as "controlling" the number of plant-eating caterpillars, are incidental.

The Evolutionary Play

And so from ecological to evolutionary time. In a few generations of interaction between a parasitic wasp and its host, both these players remain more or less the same, even though their numbers may fluctuate wildly. But in thousands or millions of generations both will change. It's sometimes called an evolutionary arms race. Or it's likened to the situation of the Red Queen in Lewis Carroll's *Through the Looking-Glass*, who has to run to stand still—species have to evolve to maintain the same fitness in the face of their enemy's evolution.

Not all evolution is like that. There's plenty of evidence that human races have evolved in response to the climate. But the climate does not evolve back. The nature of evolution is rather different according to whether the source of mortality that causes natural selection is biotic or abiotic. And in some instances of evolution, the specifics of the environment may play only a minor role. When early fish evolved jaws from a starting point of a sucker-like mouth, this can be thought of as a generalized improvement rather than as an adaptation to eating a particular type of prey. Having a more efficient prey-catching apparatus is probably beneficial to survival in all environments, whatever the exact diet. But, that said, the idea of an ecological theater in which evolutionary plays takes place is a good one. We should refine it, not reject it.

Here's an important refinement, especially for the longest of plays. The reason a trait of an animal evolves—what you might call its adaptive value—may change over time. An example is provided by feathers, those characteristic structures of birds (and, as we now know, certain dinosaurs too). It's thought that they first evolved for heat insulation and only later for flight. In the long term, many structures in many organisms have been subject to this process of shifting reasons for their elaboration. The ear bones of mammals initially evolved as part of the jaw hinge mechanism of our reptilian ancestors. And perhaps some of our most uniquely human characteristics have also evolved through the same sort of process. Let's think about this possibility.

Ancient Africa

When I first heard it, I found the ecclesiastical title Primate of All Ireland hilarious. Since the Primate—essentially another name for archbishop—is based in the ancient city of Armagh, I pictured a sort of King Kong figure living there and giving a sermon in a splendid cathedral. In fact, there are two relevant cathedrals and two such Primates, only one of which is necessarily a "him." This strange situation arises from the fact that both the Roman Catholic Church and the Church of Ireland (Anglican, Episcopalian) have such a Primate. Anyhow, as a non-religious biology student whose Presbyterian upbringing didn't include higher tiers of church officials, I came across primates before Primates; and this, of course, is the correct evolutionary order, too.

So we'll switch now from Primates to primates. We all have an instinctive view of the latter. We think of them as mammals that are either monkeys or something closely related to monkeys—for example, apes, lemurs, or bush babies. And humans of course; after all, Primates are primates as well as being clerics.

Is there a common thread that connects all the various types of primates? There are several candidates, but the best one, at least in my view, is adaptation to an arboreal existence, that is, life in the trees. The earliest primate fossils date from about 60 million years ago—not long after the extinction of the dinosaurs. Reconstructions of these animals show them as creatures that look like a cross between a monkey and a fox. We need to take such reconstructions with a pinch of salt, of course, but it's clear that the first primates could climb trees. And pretty much all primates since have climbed trees—though the group containing humans and the other great apes is less arboreal than others.

Moving through the 3-D environment of trees rather than on its 2-D equivalent, the ground, turned our front feet into hands, evolutionarily speaking, and indeed made our hind feet more handlike than those of our ground-dwelling mammalian ancestors or cousins. For example, think of the contrast between the feet of a monkey and those of a horse. And in our newfound 3-D home, depth perception was crucial—if our primate ancestors misjudged the distance between branches high in the canopy and then attempted to jump from one to the other, they would have fallen to the ground and their genes would have fallen into evolutionary oblivion. So our eyes shifted from the sides of our head to the front, giving us better stereoscopic vision—much as eyes in the owl lineage have done in the course of bird evolution.

Ancient Africa, and in particular sub-Saharan Africa about 10 million years ago, was a place of tropical forests. Many primates were evolving in these, including the ancestors of modern humans, who would shortly diverge from the lineage leading to chimps. All of us had by that time refined our various adaptations to tree living. There we were with our versatile hands, our front-facing eyes, and our enhanced brains that we used to operate these things—agile actors able to swing about in our leafy theater with rarely a misjudgment leading to a fall. And then the forest shrank. Not so much that it couldn't be lived in anymore; in fact, most primates stayed there. But at least one didn't.

It's ironic that our ancestors, brilliantly adapted for life in the trees, came down to earth again and began to walk on the ground. From the point of view of our legs, this could be thought of as adaptation in reverse. Let's compare this with the evolution of feathers.

To make an abstract version of the case of the feather, we could say that it evolved first to do X and then to do Y. In other words, temperature regulation (X) is different from flight (Y) but is not in any sense the antithesis of flight. In fact, many feathers probably help with both jobs. But our early primate ancestors' legs and feet evolved to do X, and then the legs and feet of their descendants, early humans, evolved to do anti-X. The optimal design of a leg for life in the trees cannot be the same as for life on the ground. Indeed, many accounts of human evolution focus on our upright, bipedal posture and form of movement.

It could be argued that ascent into the trees and, about 50 million years later, descent back to the ground was an essential pair of transitions without which natural selection could not have produced humans. This hypothesis leads to all sorts of interesting thoughts. Is dolphin intelligence, great as it is, limited by the fact that their ancestors were not arboreal? On distant planets, does the level of intelligence and manual dexterity associated with a civilization that can direct its telescopes out at space require an ancestry of tree dwellers that returned to a surface existence? Is evolution elsewhere in the universe even similar enough to its counterpart here on Earth that it produces the type of large plants that we call trees? Endless fascinating questions, without, as yet, any clear answers. But one thing I'd put money on: whatever evolutionary plays are going on "out there" surely take place in ecological theaters, however similar or different they are to our own, and however similar or different the earthly and alien players.

Ecological Origins

Our own origin as humans is one thing; the origin of ecosystems is quite another. But the former depended on the latter. If ecological theaters had not originated, we would not have been able to do so either. And the origin of an ecosystem is quite a different process from the origin of its constituent creatures. To see this more clearly, let's look at volcanic islands.

Islands form in various ways. Some, such as Ireland and Britain, started life as part of the continental mainland and then split off from it due to a rise in sea level turning what were terrestrial valleys into marine ones. Others were never part of a mainland. They emerged upward from the sea rather than sideways from a continent. Examples are the islands of the Hawaiian and Galapagos archipelagos.

Let's think about a single volcanic island rather than a whole chain of them. Here's a simplified but true-in-essence scenario. A volcano erupts under the sea. Magma pours out, cools, and builds up a rocky structure on the seabed that wasn't there before. In some cases this structure grows until its top breaks the surface of the water. Suddenly the Earth has a new island—one that would not have been marked on older maps because it simply wasn't there. It cools further and, in the absence of continued nearby volcanic activity, eventually equilibrates in temperature with its surroundings. At this stage it's potentially habitable but uninhabited. The only animals that can reach it easily are marine ones, and these will do no more than colonize thin strips of newly created shoreline.

It takes a very long time for a mature ecosystem to develop on our new island. Spores, seeds, birds, and insects may reach it from the skies; and occasionally various creatures may reach it on floating logs. However, most of these will be unable to flourish on the sterile rocky surfaces on which they have been unfortunate enough to land. But eventually the rocks will become partly coated with algae and mosses. And these will cause little bits of rock to crumble, forming the beginning of a gritty soil. In time, some other plants will be able to survive, and with them some insects or other small animals. It's a slow process. But if we could make a return visit in 50,000 years, we'd probably find an island covered in forest—much like the one we considered earlier with its several species of trees and associated ground flora and fauna.

The process through which an ecosystem comes into being on a previously barren piece of land is called ecological succession. We don't often get the chance to observe new volcanic islands, but the same sort of ecological process takes an abandoned quarry from bare rock to forest. In both of these cases we talk about *primary* succession because the rock surfaces that end up being colonized have not been inhabited before. Ecological succession in areas that have previously been inhabited is called *secondary.* An example of this secondary process is the development of forest on an abandoned piece of farmland.

The end point of succession depends on where we are in the world—rainforest will never be the end point of ecological succession in the Arctic, for example, and tundra will never be the end point in the tropics. However, the end point is often the same regardless of whether the process is primary or secondary. An abandoned quarry and an abandoned farm near to each other in Pennsylvania, for example, will both end up as deciduous forest—it will just take the quarry a bit longer to get there.

The process of succession toward a predictable end point, such as deciduous forest in Pennsylvania or France, looks like goal-directed behavior. It looks as if the series of ecosystems that follow each other represent an *attempt* by the collective flora and fauna to reach a sort of "adult" ecosystem. But such an interpretation should be resisted because the plants and animals are not trying to do anything except survive. What happens is that at each stage in the process of succession the dominant plants alter the environment in such a way that other types of plant that could not previously survive now can do so. Eventually, ecosystems dominated by bushy plants give way to forest—the taller trees overshade the bushes, many species of which then disappear. The trees of the forest cannot be overshaded by anything, so they persist indefinitely.

The "adult" ecosystem in a particular area is simply an inevitability of the rules of survival in a changing environment. Like the successional ecosystems that preceded it, it has no function in the normal sense of that word. This is in contrast to an adult animal (or plant), which has a very definite function: to reproduce. And that's where we're going next—the development of adult humans and other animals.

Chapter Twelve

Becoming an Adult

From Mulberry to Tenth Birthday

We looked at human mulberries in Chapter 3, where we saw that these early-stage embryos lack nerves. They also lack most other structures that we associate with humans, such as muscles, lungs, and heart. Between the mulberry stage at one to two weeks and the fetus at two to three months, all these structures, and many others, appear. This period of time is the most creative of all in human development. The creature present at its start, the mulberry, could barely be distinguished, by ordinary microscopic examination, from its equivalent in a horse or a dog, let alone a chimp. The creature present at its end, although it still has much developing to do, is a tiny human for sure, albeit one that has probably not yet had a single thought pass through its rudimentary brain.

As early fetuses, we're all about the size of the terminal section of our adult thumb. Our head takes up about half of our body length. Our individual developmental journeys now continue, and this process can be thought of as a combination of three sorts of change. First, growth: we elongate by a factor of about 30 times prior to being born. Second, change in shape: our heads become an ever smaller fraction of our bodies, a process that continues for many years after birth. Third, continued elaboration of organs: our brains develop more cells, a greater number of different types of cells, and more interconnections between cells. The brain we are born with may be like a book without any words, but at least most of the pages are there, ready to be written on, though we shouldn't take the book analogy too far—I can't change the size of a page by writing on it, but some elements of the brain can indeed be expanded by experiencing things in the post-birth world of myriad stimuli.

From birth to our tenth birthday a lot of amazing things happen, and yet in a sense we remain more or less the same in our overall form. For example, we can only squirm at first; then we can crawl, and later walk. We also become able to run, jump, and swim. But through all these extensions in our mobility, our limbs remain much the same. They grow; their muscles get stronger; their dexterity increases. But the number of fingers and toes remains the same, as does the pattern of the long-bone supports for the arms and legs—one bone followed by two in each, with a kneecap at the joint in one case but no elbow cap at the joint in the other.

One way of looking at the difference between our development as early embryos and our development as children is that in the first of these two phases new organs and structures get made de novo, whereas in the second phase they just grow—albeit at different rates, hence our altered shapes. No significant new structures appear during this decade of childhood.

The Origins of Sex

In the living world as a whole, many creatures reproduce asexually. But in the animal kingdom, asexual reproduction is relatively rare. So an interesting evolutionary question is how sex arose and persisted. There are several books about this if you're interested. Here, though, I want to focus on the origin of sex in an individual on a timescale of months or years rather than the origin of sex in the living world in general on a timescale of millions of years.

But even restricting our attention to an individual human, there are at least three origins of sex. We've already looked at the first: the inheritance from our father of an X chromosome, in which case we're a daughter, or a Y chromosome, in which case we're a son. So even as a fertilized egg we have a sex. But it's a sort of theoretical sex—just an essence, if you like. The second origin of sex starts about a month or two later. An embryonic structure called the *genital tubercle* is formed (*tubercle* is really just a fancy word for a lump). This structure is initially similar in male and female embryos; later, at about two months post-fertilization, it develops into a rudimentary penis or clitoris. Other reproductive structures begin to form at about the same time. The essence of sex has now become the structure of sex.

At this point all goes quiet on the sexual front. Not completely silent—sexual structures continue to develop slowly, like all our other embryonic structures. But quiet in the sense that the sex organs of a three-month embryo are nevertheless functionless. And a decade later, as we look forward to our first double-figure birthday, they remain unable to function. This is very unusual, in fact unique, as organs go. To have a functionless heart, liver, or lungs at age 10 would lead to pre-adult death. But nonfunctional testes or ovaries are the norm at this age.

Now comes our third origin of sex: the development of functionality in our ovaries or testes, and the development of breasts and beards. Breasts are clearly of some use in the reproductive process, while beards are not. But both are an integral part of the blast of developmental activity that we refer to as puberty, which also involves a host of other changes, such as those in body shape and hairiness of the legs.

Although perhaps "blast" is a little overdramatic, it's true that puberty is a second creative phase in our development, the first being early embryogenesis. The decade or so in between has been a period for growth rather than creativity. Our 30-fold increase in length from early fetus to newborn infant is followed by another such increase from birth to puberty. Arithmetically, the post-embryonic increase is bigger: instead of elongating by a foot we elongate by more than a meter (with apologies for the mixed measurements). But geometrically, the post-embryonic increase is smaller than the embryonic one: our length increases by a multiple of about 4, not 30 as earlier. And if we consider body volume rather than length, the equivalent multiples are 64 times (post-embryonic) versus 27,000 times (embryonic), an even starker contrast.

The Onslaught of Puberty

The most difficult years for a healthy human to cope with are undoubtedly those of puberty. At 10, most of us are happy children; at 20, most of us are reasonably happy adults. But the transition is in most cases a painful one, psychologically speaking. I'm no psychologist, so to learn more about that aspect of puberty you'll need to consult one of the many books on the subject written by an expert in the field. What we'll focus on here is the question of what kicks puberty off. So, we pose the question: what exactly is it that says to the body of the growing child that the time has now come to enter another creative phase? This creativity is not an alternative to growth, needless to say; rather, it's a supplement. Growth continues, and indeed speeds up for a while, before later slowing down and eventually, after puberty, ceasing altogether.

As we know, it's all to do with hormones. But that answer raises another question, and perhaps a more interesting one: what starts the flow of hormones? Think about it: one day our bodies are functioning normally as we run around playing the games of childhood. The next day they're doing the same; and so on. We might be age 8, 9, or 10; it doesn't make much difference. But one day something happens that changes our lives forever.

Of course, "one day" is poetic license. The origins of puberty can be traced back into the single-figure ages of seven to nine, and some biologists would say even earlier. The physically apparent bodily changes of puberty take a few years from start to finish—exactly how many years depends on how you define puberty, and also on the person, as some people make the transition faster than others. The period from 12 to 15 encompasses the transition from infertile to fertile in most developing humans; 15 to 18 is more like the icing on the cake for the body, though for mental adjustment to adulthood it's very important. And then the body settles down to its post-puberty plateau—we've stopped growing upward, though not necessarily outward. The developmental system has done its job.

Whether it's the poet's single day or the scientist's three years, the question remains: what starts the flow of hormones? We don't yet have a satisfactory answer, though in a way that's not surprising. All developmental events have causes that involve earlier developmental events, and so there's always a feeling of infinite regress in developmental biology. This ultimately leads back to the fertilized egg, but it doesn't even stop there because it continues back through a few generations as the chicken-and-egg problem; it then extends back through many more generations as it transforms into the evolutionary issue of the origin of our species.

However, we do know something about the changes in hormone production that cause our bodies to begin to enter puberty. Here's a small part of what happens. At the age of about 11 (maybe a bit earlier in girls than boys) sex hormones start to be secreted. These are primarily estrogens in girls and androgens in boys. Over the course of puberty they are secreted from more than one organ. For example, ovaries, adrenal glands, and several other sources produce the estrogen called estradiol, which is sometimes referred to as the primary female sex hormone, its equivalent in this respect in males being testosterone.

The brain is also involved in secreting hormones that play a role in puberty. Especially important are the parts of the brain called the hypothalamus and the pituitary, which are situated close to each other underneath the thinking part of the brain, the cerebrum. When a certain hormone released from the hypothalamus reaches the pituitary it causes cells there to produce a different hormone, which in turn, when it reaches the gonads (testicles or ovaries) and the adrenal glands, causes the production of yet another hormone. And in fact there is not just one such hormone in each case, there are several.

Different hormones have different roles in puberty, especially in the case of differences between boys and girls. For example, estradiol is responsible for the growth and development of the breasts; testosterone is secreted in the testes and is responsible for their development. However, testosterone is secreted by ovaries too and is involved in clitoral development, among other things. The whole system of interaction between brain and gonad hormones is called the hypothalamic-pituitary-gonad axis, and its overall operation is fiendishly complex, involving not just on / off switches but also finer controls of hormone levels and feedback mechanisms.

The main fault of my brief account of hormones and their roles in puberty is not the lack of detail; this lack can be rectified by reading a different type of book. Rather, the main fault lies in my not providing an answer to the question we started with: what *initiates* puberty? How does a complex system of interaction between multiple hormones start? And *where* does it start? Of the multiple glands involved, does it start at the base of the brain, or in the abdomen, deep within an adrenal gland or a gonad? Does it even have a "start" at all? Biologists will eventually answer all of these questions except the last, which is perhaps more within the realm of the philosopher. But we've a long way to go yet.

Mammalian Cousins

We've already seen that most male birds don't have a penis. And of course female birds lack breasts. Mammary glands and mammals go together, as their names suggest. So adulthood means different things to different creatures; but perhaps *becoming an adult* is broadly similar across all 5,000+ species of mammal? In terms of ovaries and testicles becoming functional, the answer is yes. No mammals are born capable of reproducing immediately, while by a certain age (though a very variable one) all species of mammal do have a reproductive capability. If they didn't, they wouldn't exist. And although hermaphrodite adults are common in many other groups of animals—snails, for example—the mammalian rule is for sexes to be separate. So all mammals become reproductively mature at some stage of their development, and in almost all cases they become a mature male *or* a mature female.

The variability among species in reaching the age of reproductive maturity broadly reflects both the gestation period and the life span; there's also a relationship with body size. Low values of all these variables tend to go together. Let's take typical human values as our starting point. We are born aged about 9 months; we can reproduce by our early teens, though most of us defer having offspring until many years later; our life span is usually less than a century, though if we're lucky not by much.

Now consider a small mammal such as a mouse. If I was reincarnated as a female mouse (unlikely, I suspect), I would be born at about 3 weeks and might live for a year. I would be an adult, in the sense of being capable of reproducing, by about two months after my birth rather than about 14 years, as was the case in my previous life as a human. At the other end of the size spectrum, if I was reincarnated as a male sperm whale (equally unlikely), I would not be born until about 15 months after I was conceived, and if I was lucky I'd live to about the same age as an average human. I would not be an adult until I was about 18 years old (though, strangely, my sister might be reproductively mature at 10).

Female whales, like female mammals in general, have mammary glands with nipples. But is this true of all mammal species? Nearly, but not quite. The duck-billed platypus is an exception. The babies of these strange creatures, when they hatch out of their eggs, do have a supply of a milk-like substance, but it's exuded from quite a large area on the underside of the mother's body. Many small glands lurk behind the skin in this area; each disgorges its contents separately, so there are many pores through which platypus milk emerges. This seems to be similar to the ancestral mammalian condition. In other words, we think that the protomammals of the past were like the platypuses of today in this respect. Other lineages of mammals evolved mammary glands and nipples, the number depending on the lineage concerned. Humans are at the low end of the scale, with two; the highest number of nipples in a mammal is around 20.

Nature is full of variation. Becoming an adult mammal is a broadly similar process regardless of species, but the details vary considerably. At adulthood, a female mammal can have no nipples, two, or many. A male mammal can have external or internal testicles—whales exemplifying the latter condition. A male mammal can mature at about the same age as a female, or a little later, or at almost twice the age of a female of the same species. Both the age of reproductive maturity and the length of the gestation period tend to scale with body size, as we have seen, but not as cleanly as you might expect. An elephant's gestation lasts longer than a sperm whale's, despite the fact that the body weight of an adult male sperm whale is several times as great as that of the elephant. But again, these are mere details. However, if we extend our comparisons to more distantly related animals, becoming an adult is a different process altogether.

Miraculous Metamorphoses

So let's now forget about mammals and think about some of our *much* more distant animal cousins: butterflies and moths. These far exceed mammals in terms of numbers of species: in contrast to the mere 5,000 to 6,000 species of mammals, there are nearly 200,000 species of Lepidoptera. And how each of these species makes its adults would stretch our credulity if we couldn't watch the process happening. But we can.

In a way, this story doesn't need to be told because we're all familiar with it. However, the external changes that turn a caterpillar into a butterfly are underlain by internal processes that are just as remarkable but not familiar at all beyond the world of professional biologists. Again, the details vary from species to species, but the essence of the process is the same. Here's what happens inside the body of a "teenage" caterpillar as it stops eating, grinds to a halt, and begins the next stage of its developmental journey.

Inside the hard casing of a chrysalis, most of the larval tissues and organs are broken down. Mostly they're not used to build the adult butterfly. Instead, a series of small, often disc-shaped bits of pre-adult tissue that were hiding inside the caterpillar now come to the fore. They increase in cell number and begin to take on more complex forms than quasi-discs. Many are arranged in pairs, with a left-hand and a right-hand disc. The overall series of disc-shaped and other-shaped bits of pre-adult tissue is scattered through the caterpillar's body, with different discs in different body segments.

At the head end are discs for mouthparts, antennae, and eyes. At the rear end there's a genital disc to make reproductive structures. In between are many other discs, including those for making wings. A butterfly's wings begin their development in the larva, with proliferation of wing-disc cells, but they undergo the most dramatic developmental changes in the chrysalis stage. When a caterpillar begins pupation the wings are still represented by disc-like bits of tissue, though these are more elaborate than they were at earlier larval stages. But when the butterfly emerges from its now-redundant pupal casing the wings are fully formed. The pupal period is characterized by frenetic developmental activity. Think of the difference between a tiny, almost colorless internal disc (or blob) of tissue and the magnificent wing of an adult swallowtail or tortoiseshell.

In Lepidoptera there are two pairs of wing discs inside the caterpillar because butterflies have two pairs of wings. But in houseflies there is just one pair of wing discs because evolution has dispensed with rear wings in their case, having miniaturized these and turned them into small flight-balancing structures that are called drumsticks because of their shapes. So a fly larva—in other words, a maggot—has smaller discs from which these structures are formed, rather than a second pair of wing discs.

Whether we're dealing with the caterpillar-butterfly metamorphosis or its less glamorous maggot-housefly equivalent, something has to start the process off, just as it has to in the puberty of humans and other mammals. And again that something is a hormone. Its short name is ecdysone and it is produced in glands at the head end of the larva. Acting in conjunction with other players in the molecular game going on inside the metamorphosing body, it acts to turn larval genes off and adult genes on. We understand quite a lot about how it does so. However, as with other developmental processes, each cause (in this case a hormone) is itself an effect of an earlier cause, all the way back to the egg. It's infinite regress again. But such regress is ultimately analyzable, whereas miracles are not. So my title for this section is just another piece of poetic license.

Adult Fates

We could say that the fate of a child is to become an adult human and that the fate of a caterpillar is to become a butterfly, though in neither case is this progression inevitable. Premature death is always waiting in the wings as an alternative fate. In modern human civilizations, the likelihood of a child becoming an adult is high. For a caterpillar, the likelihood of it metamorphosing into a butterfly is rather low.

But what is the fate of *adults,* be they humans, butterflies, or other kinds of animals? Ultimately death, of course. Of the two supposed certainties in life, death and taxes, only the former is truly certain—with apologies to Benjamin Franklin. Some primitive human societies have no taxes, and butterflies are blissfully unaware of the subject.

Let's try to come up with a more positive answer to the question of adult fate than simply death. If death were the *only* fate of adults, the living world would not exist. At the very least, *some* adults must reproduce as well as die. And although humans are an exception, most animals reproduce in a seasonal pattern. So there's a cyclicity to reproduction, in contrast to the straight line through time leading to death.

There are some great metaphors for the dual nature of time, as faced by the inhabitants of planet Earth, be they humans or other creatures. The American paleontologist Stephen Jay Gould wrote a book called *Time's Arrow, Time's Cycle,* which captures the duality beautifully. And if we try to hybridize an arrow and a cycle we get a helix, or coil. Hamlet's reference to "shuffling off this mortal coil" springs to mind at this point. The adults of all kinds of animal, not just humans, have a mortal coil, during which they reproduce a variable number of times and eventually die.

What's the cause of the seasonal cycles that we all experience? It's the motion of the Earth around the Sun, coupled with the Earth's significant tilt. This combination gives rise to many cyclic patterns in the lives of humans and other animals. The details of the patterns and their exact causes depend on where we live. In the temperate zone, seasonal variation in temperature is pronounced. In many tropical regions, seasonal variation in rainfall is more prominent. We usually take Earth's seasonality for granted. But we shouldn't be satisfied with such a lazy mental stance. So let's now have a look at how the whole system of cycles and seasons originated, billions of years ago.

FROM BOULDERS TO BRAINS

Chapter Thirteen

Rubble around the Sun

Before the Sun

We'll begin our exploration of the origin of the Earth and its seasons by making two impossible journeys, one in space and one in time.

We climb aboard a gleaming new space-time machine, close and seal the entry port, engage the engines, and launch ourselves into space, in the direction of the center of the Sun. The distance we have to cover is only about three times the distance from Earth to Mars when these two planets are closest together; with current technology that's doable in a few years, and with our hypothetical space-time machine probably much faster. The reason the journey is impossible is not its length but rather the conditions we'd encounter in its later stages. When we reach the surface of the Sun the temperature has risen to more than 5,000 kelvin, but that's cool compared with the temperature when we reach the Sun's core: about 15 million.

Suppose that we and our craft survive, protected by a thick outer shell of insulating material, which keeps the craft's internal temperature the same as that of a Spanish summer. We now adjust the controls so that our craft switches from space travel to time travel; we travel back in time by 5 billion years and then stop. We don our space suits and get out of our craft, tethered to it by space ropes. It's not too hot now because the Sun doesn't yet exist. We look around in all directions: what do we see?

The short answer to this question is very little. Five billion years ago, the space that would later be occupied by the Sun looks rather boring. We see no nearby suns or planets. Perhaps in the distance we can see some stars, but they look a bit hazy. We're in the middle of a large cloud of gas, dense enough to look misty but sparse enough to allow us to see some of the brightest stars through it. The gas consists mostly of hydrogen but with small amounts of other elements. It's definitely not a breathable atmosphere. There's no sign of anything happening, no indication of things to come. We shrug, climb back into our craft, and return home.

By 4.6 billion years ago, not long after our visit in the grand scheme of things, the gas cloud has collapsed under its own gravity and become incredibly dense and hot, so hot that nuclear fusion of hydrogen to helium has begun in its core. The gas cloud has become the infant Sun. But what of the Earth and the other planets?

Nothing in space is neat and tidy. Orbits are not perfect circles. Neither stars nor planets are perfect spheres. Spiral galaxies are not perfect spirals, whatever we deem those to be. The plane of our galaxy isn't perfectly flat—it's warped. Earthly mathematics simplifies complex shapes to perfect ones, since that's a great way to make progress in understanding things. But we shouldn't confuse models of reality with reality itself. Our solar system is messy now, in ways that we'll get to soon; and it was even messier at the time of its origin. Our job in this chapter is to look for patterns among the mess. The present-day patterns are quite easy to discern. Their ancient counterparts are harder but not impossible.

From Dust to Rubble

About 4.6 billion years ago, the collapsing cloud that gave rise to our solar system looked like a disc of hazy material with a bulge in the middle. This bulge was the proto-Sun, while the flatter outer region consisted of a lot of gas and dust that would soon coalesce in various ways to form planets. This outer region is called a protoplanetary disc.

The most important transition to occur in the proto-Sun was crossing the temperature threshold that enabled nuclear fusion to begin. The most important transitions (plural this time) to occur in the disc were the collisions and adherences of small particles, thus creating larger ones. When the disc first formed, it probably consisted mostly of gas molecules, but also some larger particles such as ice crystals and dust grains. But exactly what is meant by dust? On hearing that word in an everyday context, we tend to think of those little specks that can be captured in a beam of sunlight from a window, which are especially numerous if children have just been having a fight using cushions as weapons. Since we can see them, these dust grains are probably about a millimeter across. Cosmic dust grains are generally smaller than household dust grains—ranging in size from just a few molecules to about a micrometer (a thousandth of a millimeter).

Dust grains in space collide with each other due to their random motions, and often, on colliding, they adhere to each other, though exactly how they adhere is not fully understood. The New Zealand planetary scientist Stuart Ross Taylor talks about "grains that stuck together by obscure processes" in his 1998 book *Destiny or Chance* (59). Whatever the mechanisms—probably a mixture of "sticky" surfaces and electrical charges—as the collisions and adherences continue, the resultant particles get bigger and bigger. Eventually they reach a size where "particle" is no longer an appropriate name. They become collectively what I've chosen to call *rubble*. But how big are individual bits of rubble, in the context of the early solar system? They're very variable in size but many of them are the size of large rocks—say about a meter in diameter. Further collisions result in even larger rocks and eventually large bodies called planetesimals, perhaps a kilometer or so across.

These bodies are large enough to attract other nearby bodies through gravitation. So, further collisions and mergers occur, leading to the formation of protoplanets of more than 100 kilometers diameter. The largest of today's asteroids, such as Ceres and Vesta, are about the same size as some of the early protoplanets. Ceres' diameter is almost 1,000 kilometers, about a quarter that of the Moon. The Moon itself is about the size of the larger protoplanets.

Catastrophic Collisions

So far, we've seen the early solar system evolve from a cloud of gas and dust to a central Sun with a lot of large bodies orbiting it. How many such bodies there were cannot be established with certainty, since most of them became obliterated or absorbed in subsequent collisions. But it's thought that there were hundreds of Moon-sized protoplanets, together with lots of smaller ones and a few bigger ones. With so many large objects hurtling around the Sun, collisions were inevitable. Mergers of protoplanets into the eight planets that we see today probably took a few million years—not long in astronomical terms. Not all of the smaller objects ended up as parts of larger ones—some still exist today, most of them either in the asteroid belt between Mars and Jupiter or in the Kuiper belt beyond Neptune. But the collective mass of these two great collections of rocks and "dirty snowballs" is just a tiny fraction of the mass of the solar system as a whole.

If we ignore this remaining rubble, our view of the solar system at the end of its formative period is of a central star, the Sun, orbited by eight quite well-spaced-out planets—four inner rocky ones and four outer giants made largely of gas (Jupiter, Saturn) or ice (Uranus, Neptune). This arrangement seems to have been reasonably stable even though thousands of further collisions have occurred. We see evidence for these in the craters that pepper the surface of the Moon.

Rotation, Rotation, Rotation

Most objects in space rotate, but *rotation* can mean at least two things: rotating on one's own axis (spinning) and rotating around something else (orbiting). Also, the concept can be applied to several levels of astronomical entities, from moons to galaxies. Here we're concerned primarily with rotation within the context of our solar system, both in the distant past and in the present. We touched upon this above when thinking about collisions between protoplanets, each of which was orbiting the Sun. As with their exact number, their exact orbits are no longer known. But the orbits of today's planets are known with an incredible degree of accuracy. Let's have a look at these orbits, and also the planetary spins.

So we'll take another trip in space—but not time. We'll launch from the North Pole and head straight up. We'll keep going until we're far enough up that we can get a plan view all the way from the Sun out to Neptune. Then we stop and look down at the whole system from our northerly vantage point to observe the planetary movements. Here's what we see.

We'll start with the Sun. It rotates on its axis counterclockwise about once a month, though because it's not solid, some bits of it (near its equator) rotate faster than others. We don't observe the Sun orbiting any other body and we pretend that it remains at a fixed point in space. That's not true, but its long-term motion relative to the galaxy can be ignored for our present purposes.

Now we scan across all eight planets, from Mercury out to Neptune, and we notice that they all orbit the Sun both in the same direction (counterclockwise) and in the same plane (the solar system is approximately flat from a planetary perspective). These interesting facts can hardly be coincidences, so it's natural to inquire about their causes. The common direction of orbit—counterclockwise—was caused by the origins of the planets from the protoplanetary disc, which was itself rotating counterclockwise when viewed from above. And the common orbital plane resulted from the fact that the original disc was indeed a disc and not a sphere.

But, as in the case of embryogenesis in animals, we can keep going back looking for earlier causes. Why was the protoplanetary disc approximately flat? And why did the original disc spin at all, let alone counterclockwise?

These are difficult questions. The answer to the first one concerns something called angular momentum. Since this is unfamiliar to most people, let's discuss pizza making instead. Imagine a pizza expert spinning a lump of dough. As the dough spins, the lump flattens out. This is what happens when any object, be it a lump of dough or a cloud of gas in space, rotates—the rotation causes the flattening.

Now we reach an even trickier problem: why was the cloud of gas and dust from which our solar system formed rotating at all? Why was it not just static, whatever "static" means in outer space? There are two ways to answer these questions. First, you could think of "static" as just one possibility out of millions—the others being various rotational speeds in both clockwise and counterclockwise directions. Thus it's far more probable for a cloud in space to be rotating than to be still. Second, such clouds will be influenced by other clouds. Indeed, most giant molecular clouds fragment as they collapse, with each fragment potentially giving rise to a star. If interactions between nearby fragments cause rotation to begin, it will never stop—unless some specific force acts to cause it to do so, which seems unlikely.

So, rotation is there from the start. Rotation causes flatness. But the direction of rotation is an accident, as clockwise and counterclockwise rotations are equally probable. In fact, if you choose to look at our solar system from high above the South Pole rather than the North Pole, its rotation is then seen as clockwise—the opposite of what we saw initially. Once rotation has started in one direction, it continues in that direction. When the disc breaks up into discrete bits (planets), they retain the rotational direction of the original disc.

Spinning and Tilting

The uniformity of the orbital directions of the planets is not matched by uniformity in their directions of spin on their own axes. There's still a preponderance of counterclockwise spins, but it's just that—a preponderance. The exceptions are Venus and Uranus. Venus spins very slowly clockwise, so slowly that one Venusian day takes 243 Earth days. And, bizarrely, a Venusian day is longer than a Venusian year, because the latter is only 225 Earth days.

Uranus is stranger still. Most of the planets have a slight tilt in that their axis of rotation is not at a right angle to their plane of orbit. This is what causes seasons. The Earth's tilt is about 23 degrees; Jupiter's tilt is a mere 3 degrees; but Uranus's tilt is more than 90 degrees, which means that effectively it's a planet lying on its side. We believe that all these tilts, and possibly Venus's backward rotation too, are due to collisions in the late stages of the solar system's formation.

It's worth probing a little further into Earth's seasonality, which is indeed caused by our planet's tilt, producing the cyclicity of our "mortal coil," as noted in Chapter 12. This form of seasonality produces a situation in which summers and winters in the Southern Hemisphere are inverted, so to speak, compared with their Northern Hemisphere equivalents. But there's another possible source of seasonality on an orbiting planet, and one that would give rise to the seasons running in parallel at all latitudes. This is the elliptical nature of orbits. Comets have extremely elliptical orbits, and their summer occurs when they're closest to the Sun. We Earth-bound creatures do not feel such "elliptical summers" because our orbit is very nearly circular. Nevertheless, our closest annual point to the Sun is about 3 million miles closer than our farthest. This does produce an effect on the climate, but one that is so slight we don't notice it.

Theoretical Theia

So far, we've concentrated on planets. But most planets have moons—what about these? There are too many moons in the solar system to examine all of them (Jupiter alone has about 70). But our own Moon and its origin are essential topics, especially with respect to terrestrial life. If we had no Moon, we would have much slighter tides, with significant consequences for tidal-zone and shallow-sea animals, which may have included the first animals of all, and thus our direct, albeit distant, ancestors (see Chapter 14).

There are various theories about how the Moon arose, but there's an increasing consensus that it was formed as a result of the collision of a protoplanet called Theia with the early Earth. Theia is thought to have been a little larger than Mars, so its impact with the Earth must have made the dinosaur-killing asteroid impact that took place much later look like the arrival of a large hailstone. Probably the Earth was severely damaged, losing large chunks of its prior self into space, and probably Theia itself was broken up into pieces. We believe that these chunks and pieces, probably in molten form due to the heat generated by the collision, re-coalesced either with the Earth or with each other, the latter coalescence forming the Moon.

Some of the larger moons belonging to other planets probably formed in the same way, through major early impacts. But most of the smaller moons, such as the two little moons of Mars, Deimos and

Phobos, are thought to have a more recent origin. They probably started off as asteroids that, through collisions with other asteroids, got displaced from their original orbital belt and were propelled close enough to Mars to get captured by its gravitational field.

It's not impossible that further new moons will arise by subsequent expulsions from the asteroid belt followed by gravitational capture by a planet, though the trajectory and speed of approach are critical, because they determine whether the displaced asteroid will collide with the planet or end up orbiting it. But aside from the possible appearance of new small moons from time to time, the solar system's structure seems quite stable now. We could almost say that it has grown up to be an adult.

A Strange Sort of Adult

So, the solar system has stabilized—for now. And we think that its adult phase will persist into the future for about as long as it has persisted in the past—at least another 4.5 billion years. When we get that far into the future, the Sun will begin to die, turning first into a red giant and then a white dwarf. If any of the planets survive these death throes of the Sun, they certainly won't be habitable. So we have about 4 billion years to find a new home. That should be long enough if we don't become extinct beforehand, which is perhaps more likely—though not inevitable. We'll come back to this issue in Chapter 20.

But the use of "adult" for a solar system is a bit odd. Such a system has no offspring. It contributes material to new solar systems, as we saw in Chapter 5, but these new systems bear no particular resemblance to their parents and are not, as far as we know, subjected to a parallel process to the Darwinian selection that acts on animals and plants.

The important relationship between solar systems and animals is not the parallel between them, which is only superficial—an early creative phase followed by a more stable "adult" phase. Rather, it's the fact that one is necessary as a home for the other. If extraterrestrial animals are ever discovered, we'd expect them to live at or near the surfaces of rocky planets like the Earth, not in the depths of interstellar space or in black holes.

We've already seen that the origin of life on Earth occurred quite soon, in geological terms, after the origin of the solar system itself. Our current estimates of the times of these two origins are about 4.5 and 3.8 billion years ago. But how long after the origin of life was the origin of animals? That's the question to which we seek an answer next.

Chapter Fourteen

The Very First Animals

Life and Breath

Animal means at least three things. To a biologist, all million-plus species of the animal kingdom, both vertebrate and invertebrate, are animals. But some people talk about "animals and birds," most of them using animal in this context to mean *mammal*. I was curious about how widespread this usage was, so I did a Google search for the phrase "animals and birds" and there were millions of hits. This is despite the fact that to a biologist the phrase "animals and birds" is a logical impossibility, since birds are animals just as much as mammals are. The third usage can be seen in the phrases appearing on the packaging of certain cosmetic products: "not tested on animals." This usage implies that humans are not animals, which to a biologist is a bit like saying that eagles are not birds. Any such product most certainly *has* been tested on animals—us.

Animal comes from a Latin source that means, depending on whom you believe, "life," "breath," "mind," "soul," or some combination of these. Life and breath we animals share with plants. Mind applies to some animals but not all, since some don't have nervous systems, even as adults. Soul, in the sense of an immortal one, is an open question, but it's doubtful if all animals (or any animals, depending on your viewpoint) have one. So none of these things uniquely defines animals and separates them from all other life-forms on planet Earth.

How can we talk about the very first animal if we don't have a clear idea of what an animal is? One way out of this problem is to think about some creatures that used to be considered to be animals and now aren't: protozoans. These are tiny single-celled creatures, the most famous of them probably being the amoeba, a squishy shape-changer that feeds by extending arms of cytoplasm around an unfortunate bacterium and engulfing it. Nowadays, one of the main criteria for animalness—but not the only one—is multicellularity. If you're multicellular, you might be an animal (or a plant or a fungus), but if you're unicellular throughout your life cycle, you're definitely not. So, to understand the essence of being an animal, we should start in the realm of unicellular ancestors.

Tiny Ancestors

We looked at the origin of life in Chapter 8 and saw there that the first life-forms were bags of big molecules that could reproduce. The bags were cell membranes, so these were unicellular creatures. Now we want to know the answer to the question: what was the route from these protocellular bags to protozoans and then to the first animals? In broad terms, though not in detailed ones, we know the answer surprisingly well. And here it is in outline.

Bags of molecules that survived and reproduced more effectively than others prevailed over those with a tendency to rupture and disintegrate. This was Darwinian selection operating on the infant Earth. As a result of this selection, the coherence of these tiny creatures improved. Sometime later, the hereditary material changed from RNA to DNA. At first the DNA was just loose, scattered among other molecules within what we may now call the cell. Then it became concentrated in the center. This stage of organization is that of a bacterium—either an ancient one or one of its present-day descendants.

Although the simple cells of bacteria have an outer membrane, they lack internal ones to separate some parts of the cell from others. More advanced unicellular creatures such as the amoeba have internal membranes too—for example, one that surrounds the genetic material in the center of the cell, thus forming a bounded nucleus.

The earliest known fossils, as we saw earlier, are about 3.7 billion years old and were primitive microbes that produced the layered structures that we call stromatolites. How long did it take, from this starting point, to evolve cells with a membrane-bounded nucleus? It seems that the answer is more than a billion years, which is a very long time even in evolutionary terms. And it must have taken longer again for these cells with nuclei (called eukaryotes) to have diversified into many ecological niches. The paleontologist Andrew Knoll, in his book *Life on a Young Planet,* refers to "the 2-billion-year ecological hegemony of the bacteria" (239).

The route from the earliest microbes to creatures like amoebae seems to have gone via a link to the group called archaea (literally "old ones"), which, like bacteria, lack nuclei. With the invention of the nucleus in what we might call proto-amoebae, the nucleated life forms diversified into many groups. The amoebae are just one such group, and are not thought to be animal ancestors. The unicellular eukaryotes that seem to be most closely related to animals are the choanoflagellates, whose name means collar-whips: at one end of the cell is a flange that looks like a collar, and from the center of it emerges a long whiplike structure that's used for locomotion, among other things.

Given that there are numerous groups of unicellular eukaryotes, why do we think that this one is the most likely candidate for animal ancestry? Because sponges, which are primitive animals, have a cell type called a collar cell or choanocyte, which looks remarkably like one of those unicellular life-forms called collar-whips. Perhaps the origin of sponges lay in the adherence of groups of these initially single-celled organisms. We'll probe this possibility further shortly. But for now we need to broaden out and get a big picture of the relationship among animals, plants, and other life-forms.

Trees in the Sea

Let's take an unusual but vivid approach to the relationships among the major groups of life-forms here on Earth. It involves several trees growing out of an expanse of ocean. Of course, trees normally grow on land; but then again, these are not normal trees.

Consider being on a small boat on the surface of an imaginary ocean. There's no land to be seen anywhere. And yet there are tree trunks emerging from the water, about 10 of them. They're widely separated from each other and of different sizes. Some are huge, like California redwoods; others are more like diminutive mountain ashes. Although they're very different sizes, they all have many branches. One of the trees is red, another green. The colors of the other eight are hard to determine because these are smaller than the first two and more distant from our vantage point.

Now here's the reality that our mental image of trees in the sea is intended to represent. The sea is composed not of water but of all the various groups of single-celled life-forms. Anything that projects up out of the sea is a multicellular creature, and the relationships of these to each other are indicated by the branches. The different unicellular groups compose different patches of sea, so the animal kingdom (the red tree) grows out of a collar-whip patch of water, while the plant kingdom (the green tree) grows out of a patch of unicellular algae.

The existence of the other trees is interesting, because it shows that not all multicellular creatures are animals or plants. One of the trees leads to brown seaweeds, which therefore are not plants—a counterintuitive conclusion since they photosynthesize, albeit using brown pigments rather than green ones. Another tree, growing out of a patch of yeasty sea, leads to mushrooms, toadstools, and their kin. This is the fungal tree. And in fact there are probably several of these trees because it's likely that multicellularity originated more than once in fungi. Although there's no bacterial tree, there are a few bacterial twigs, since these normally unicellular creatures are sometimes found in the form of strings or mats of cells sticking together.

Our job now is to examine how the red tree grew out of the collar-whip patch of ocean. To do that, we need to examine the earliest animal fossils. The trouble is, though, that it's very hard to tell whether some of the earliest fossil "animals" were animals at all. So we'll start with *quite* early animal fossils, which are definitely the lithified remains of ancient animals, and then go backward in time to fossils whose nature is harder to discern.

An Enigmatic Explosion

From about 540 million years ago, there were animals all over the place. They were probably all marine creatures; certainly none of them walked on land. They belonged to a mixture of present-day groups. Between 540 and 500 million years ago there were arthropods, molluscs, and even primitive fishlike vertebrates. Their fossils are found in several parts of the world, including Canada (the famous Burgess Shale area in British Columbia) and China (the almost equally famous Chengjiang fossils, from Yunnan province). This period of time, officially ending 485 million years ago, is called the Cambrian, and the sudden appearance of all these animal fossils not long after the start of this period is referred to as the Cambrian explosion.

If we go back just a little, to 550 million years ago, the typical Cambrian-type fossils were entirely absent. There are two interpretations of this stark contrast. One is that the Cambrian explosion was an evolutionary phenomenon—a sudden and dramatic branching of animal lineages following the origin of the first animal. The other is that it was merely an explosion of fossilization, in which case animals were already diverse toward the end of pre-Cambrian times but left no fossil traces, perhaps because they lacked hard parts such as shells and bones.

What kinds of animal-like fossils do we find 550 million years ago and earlier? Most of them belong to a group called the Ediacaran biota, named after the Ediacara Hills, in South Australia, where they were first found. These fossils take a variety of forms, most of them inscrutable. They're large—up to a meter or so in some cases—so they were clearly multicellular creatures. Some look a bit like animals (segmented worms) and some look more like plants (frondlike structures). Views on what these creatures were vary widely. Personally I support the view that most of them were neither animals nor plants but rather belonged to another multicellular kingdom—another tree emerging from the ocean of unicells, but one whose members are now all extinct.

Given that some other scientists think they were animals, the Ediacarans are either an interesting clue to animal ancestry or the proverbial red herring. But in either case we can look back in time earlier than them and see what fossils we find.

The Ediacaran period began about 635 million years ago, at the end of an ice age called the Marinoan glaciation. This ice age lasted for about 15 million years. So what we'd really like to know is what (if any) animal-like creatures lived before about 650 million years ago. Sadly, we don't know, though we do have some ideas and some (contentious) evidence. In particular, there are claimed Australian fossil sponges from this time, but they're small, only about a centimeter across, and their interpretation is problematic. We'll return to sponges shortly.

Fossils versus Molecules

The fossil record seems to be telling us that there were many animals in the sea 540 million years ago. Earlier than this, about 600 million years ago, there were many marine creatures but some, all, or none of them were animals—we really don't know for sure. Earlier again, at about the 650-million-year marker, there *may* have been animals similar to the simplest of today's ones—sponges. Perhaps, then, the very first animal lived about 700 million years ago? Maybe, but another line of evidence suggests that it lived twice as long ago as that.

In the late 1980s and 1990s, a new line of approach to the timeframe for the first animal began to be pursued. It's more abstract than fossils, which are the ultimate concrete evidence. So it's harder to explain. But here's the essence of the method concerned.

You take a sequence of DNA, one that represents a particular gene that's present in most or all animals. You pick an animal, let's say a chimpanzee, and you determine the exact sequence of the gene's several thousand building blocks, a sequence that's different from those of all other genes in the chimp. Then you repeat this sequencing exercise for the same gene in a different animal—say, a salmon. From the fossil record you know roughly when chimps and salmon diverged from each other: about 400 million years ago. So you can work out the rate of evolutionary divergence in the DNA sequence of this gene, in terms of the percentage difference that accumulates, on average, every million years since the last common ancestor.

Now you stick with the chimp but ditch the salmon. Instead you pick an animal that's about as distantly related to a chimp as it's possible to be. That takes us to sponges. So you take a species of sponge and again examine the DNA sequence of the same gene. Amazingly, though many genes in the chimp (e.g., those for the blood protein hemoglobin) have no equivalents in the sponge, many other chimp genes do. This time you know the percentage difference between the chimp and sponge genes but you don't know the divergence time of their lineages because the early fossil record is too poor. However, you can use your earlier estimate of the rate of divergence per million years to estimate when chimp and sponge lineages diverged.

The results that biologists undertaking the pioneering studies of this kind obtained were varied, but most of them indicated a much earlier divergence than the fossil record suggests. Some of them even suggested that the age of the animal kingdom was approximately double the fossil-based estimate, meaning that the first animal lived about 1.4 billion years ago. This is possible but, at least in my view, highly unlikely. Anyhow, later refinements of the DNA-based approach to the era of the first animals have produced estimates of the age of the animal kingdom that are closer to those of the fossils.

So, here's the current position. There were lots of disparate types of animals in the Earth's oceans 500 million years ago. It seems unlikely that all these types arose in about 2 or 3 million years around the beginning of the Cambrian Period. How long it really took to get from the very first animal to the Cambrian explosion, nobody knows. It could yet be as little as 50 million years or as much as 500 million. But despite the remaining uncertainty on this issue of timing, there is a consensus view that the earliest animals were similar to today's sponges.

The Primordial Sponge

The little collar-whip creatures that we believe were the ancestors of animals moved around and captured food by wiggling their whiplike flagella. Sponges don't move around, at least as adults; the equivalent cells in sponges wiggle their whips for feeding rather than wandering—the wiggling creates water currents that move through the sponge's hollow body, bringing with them both tiny creatures and detritus, which the sponge then consumes and digests.

So, despite the almost plantlike growth form of a sponge, from the point of view of a little marine organism caught in a sponge-induced water current, the sponge is a huge voracious predator.

In a modern sponge there are about 10 types of cell, of which the collar cell is one. The others do a variety of different jobs; for example, sponges, like us, have sperm and egg cells for reproduction. Ten cell types is a tiny number compared to the mammalian equivalent of about 200. And the sponges of the past probably had even fewer—maybe five or less. The very first protosponge was probably just an aggregation of collar cells. All it takes for such an aggregation to occur is for the cells to stick together in clumps, and all that takes, in turn, is the production of a sticky substance. Genes that make sticky proteins have recently been found in single-celled organisms. So we have a reasonably complete hypothesis as to how the very first animals arose from unicellular collar-whips, even if the timing is still a bit obscure.

None of the approximately 10 cell types of a sponge are nerve cells. So sponges are unusual among the major groups of today's animals in having no nervous system. The evolutionary route from a spongelike ancestor to humans involved three stages, as far as nerves are concerned. The first was the origin of this cell type and the formation of a basic network of these cells, capable of conducting messages. The second was the concentration of a large number of nerve cells in the head (though bear in mind that a head had to be evolved too—sponges don't have heads). The third was the elaboration of this protobrain into a huge collection of many billions of cells, a megabrain capable of ultimately thinking in abstract terms about its own origins.

As in other cases where we talk of "stages" in a continuous process, this is a mental convenience rather than a statement that there are abrupt switches from one stage to another in reality. But we shouldn't underestimate the importance of such conveniences as aids to understanding.

Regardless of whether we think of nervous system evolution in the lineage leading to humans as a series of stages or a continuous process, the development of an individual human being retraces this sequence in less than 20 years, compared to its evolutionary equivalent of about 700 million. And that's where we're going next: the embryonic (and post-embryonic) development of the human nervous system, and in particular the brain.

Chapter Fifteen

Here Comes the Brain

Sausages Revisited

We don't have to start this story—of the development of our brains—at the very beginning, because it began in Chapter 3. But since much of interest has intruded in the meantime, let's refresh our memories in terms of the very first steps. We start as a fertilized egg—no sign of nerves there. The egg multiplies up to about 100 cells in the form of a ball, the mulberry—again, no nerve cells are to be seen. Then the ball undergoes further cell proliferation and complex changes in shape, so it ends up elongate, like a sausage, with one end as the head and the other as the tail. Running along the dorsal midline of the sausage is the forerunner of our central nervous system. It starts as a thickened plate of cells running from head end to tail end. Then the edges of the plate bend upward, meet, and fuse, forming a tube. This structure, the neural tube, will become the brain at the head end and the spinal cord along the rest of the body.

So, having ended with sausages in Chapter 3, we resume with them here. Let's try to get a more complete mental image of the sausage that was you or me when we were tiny embryos. A cross section of the sausage would be approximately circular. Near the top, and just a little inside the skin, the neural tube would be approximately circular too—a circle within a circle. Bringing back the third dimension, this translates into a tube within a tube. And this inner tube is the part of the human embryo whose further development we're going to trace. It's worth bearing in mind, though, that the rest of the embryo experiences the development of nerves too. We call these collectively the peripheral nervous system to distinguish them from the brain and spinal cord, which together comprise the central nervous system.

The spinal cord's development is much simpler than that of the brain—with the proviso that *all* developmental processes, when looked at in detail, are astoundingly complex. The cord increases in diameter and complexity. In particular, the fates of dorsal (back) and ventral (front) cells are different. Sensory nerves develop dorsally and motor nerves ventrally, so that when the cord has developed to the point where reflex reactions occur, any stimulus, for example heat applied to the hand, comes in at the back of the cord and then nerve impulses exit at the front. This process leads to muscle contractions that remove the hand from the heat source.

Such reflex reactions do not involve the brain at all. If they did, the response would take longer, with consequent detriment to our health, because the hand would remain in contact with the heat source for longer and thus incur a more severe burn. But most other nervous system activities *do* involve the brain; activities such as reading and writing would be impossible without it. So now our question becomes: how does the head end of the early embryo's neural tube develop into a complex thinking machine consisting of perhaps 100 billion cells? As questions go, this one's pretty tough, but we won't let that stop us from trying to answer it.

Bulges and Splits

The head end of the neural tube, the bit that will become the brain, grows thicker than the rest of the tube, the long posterior stretch of it that will become the spinal cord. But this anterior thickening is not a uniform process. Rather, what happens is that three bulges develop: the forebrain, midbrain, and hindbrain. As well as being different bulges, these have different fates. The embryonic forebrain will give rise to the thinking part of the adult brain, the cerebrum, consisting of the left and right cerebral hemispheres. The embryonic midbrain will give rise to the cerebral aqueduct, part of the internal cavity system in the brain, which contains the cerebrospinal fluid. The hindbrain will give rise to the cerebellum; this word means "minibrain," and its use reflects the fact that the cerebellum looks like a miniature version of the cerebrum. One of its main functions is the control of movement and posture.

In each of the three cases—fore-, mid-, and hindbrain—multiple structures are developed rather than just the single ones mentioned above. And, as part of this multiple developmental process, both the forebrain and the hindbrain (but not the midbrain) split into two parts at an early stage. However, we're walking the usual tightrope between too much and too little detail, so let's ignore most of these complexities and the jargon that goes with them. In fact, we'll concentrate on the development of the main thinking part of the brain, the cerebrum. But before we narrow down, we'll spread out.

Brains, Hearts, and Computers

A brain has one thing in common with a heart and one with a computer. Its commonality with the heart is that it arises from a developmental process, while its commonality with a computer lies in its being a complex system of electrical impulses that permit various mental tasks to be undertaken, such as mathematical calculations. Looked at the other way round, a brain's major difference from a heart lies in its mode of functioning, while its major difference from a computer lies in its mode of origin.

Given that computers are human artifacts, it's clear that we understand both how they are made and how they function, though in this case it's a fairly selective subset of humans that constitutes the "we," and one that, regrettably, doesn't include me. But when it comes to our hearts and brains, our understanding is very incomplete, even when "our" encompasses every human on the planet, including developmental biologists.

The heart's function is easier to understand than its development. Underneath all the complexities, the heart is a pump. It propels blood through the labyrinthine system of vessels that permeates our bodies. It's a pump that has to keep running without the slightest break for maintenance over many decades. In most people it does just that—though, as ever with both biological and human-made machines, it occasionally becomes faulty, usually with dire consequences.

The development of a human heart from an embryo that lacks blood cells just as emphatically as it lacks nerve cells is much more complex than the heart's function but much less complex than the development of the brain. Most of the heart is made of muscle cells, albeit of a special type only found in cardiac muscles. Both cardiac and ordinary (e.g., biceps) muscle cells work by contraction. The biceps is a simpler case because when it's bending the elbow, it's essentially working in a plane of two dimensions. Heart muscles are distorting the shape of a hollow structure, and hence are working in three dimensions. But the actual contractions are similar in that they involve strands of two contractile proteins. For these to do their job properly, the protein fibers and the cells that they're in must be running in the right direction, which, with a few complications, means parallel to each other so that they all pull the same way and thus complement each other's contractions rather than canceling them out.

The problem with brains is that in order for them to think properly (or at all), the various nerve cells have to be connected up in very specific patterns—more like the electrical circuitry in a computer than the muscle fibers of a heart. This brain circuitry is more complex than that of a computer or the structure of a heart by a very long way. So we have the worst possible combination to try to understand: incredibly complex structure and function produced by incredibly complex development. Just as well, then, that we have at our disposal to approach these issues the very organ concerned.

One of the most important features of the human brain is the pattern of connections among nerve cells, and especially those in the cerebral hemispheres. We need to try to understand how one actual pattern out of many possible patterns comes about. But we're not yet ready to address this question because the point we got to earlier in this story, before our digression into hearts and computers, was a slightly swollen head end of the embryonic neural tube, with three distinct bulges representing the fore-, mid-, and hindbrain. It's a long journey from there to the brain we have at birth, and it's a long way further until we have the adult brain that we use to deal with information about our own ability to think.

The White and the Gray

It's easier to think about how to develop something if you know what that something is at the outset. So let's now look at the broad structure of our brains and then backtrack to see how adult structure arises from the starting point of three little bulges in an embryo.

It's well known that our brains (and indeed our spinal cords too) are divided up into white matter and gray matter. Crosswords and Sudoku are often referred to as puzzles that tax the gray matter, implying that the gray matter is doing the thinking. Up to a point that's true, although not much thinking would get done successfully in the absence of the white matter.

The gray matter, generally located on the outside of our cerebral hemispheres, is mostly composed of cell bodies. While we're still alive it appears only gray*ish,* with tinges of other colors, such as pink, from the blood vessels that run all over it and through it. The white matter mostly consists of long thin outgrowths from nerve cell bodies that are called axons. These are wrapped in sheaths of a fatty substance called myelin, which is what gives the white color. It's interesting that one of the main components of myelin is cholesterol—a substance with a bad reputation in other contexts but a good reputation here. Myelin can be thought of as being like the plastic sheaths that surround electric wires underneath their outer cable. Without these, we'd blow a fuse as soon as we turned on the appliance concerned. Nerves in which the myelin sheaths become damaged also don't work properly. Such damage is the cause of several diseases, including multiple sclerosis.

Overall, the nervous system is made up of two main classes of cells—neurons (commonly just called nerve cells) and glial cells (sometimes just called glia). To a rough approximation, the neurons make up the circuitry and the glial cells provide the insulation, or myelination. Both classes of cells are developed from neural stem cells. There's something of a time trend during embryogenesis such that neurons start to get made a bit before glial cells do. And both types of cells continue to be made after embryogenesis—we'll come back to this later.

But the development of the brain is far more than just the differentiation of neurons and glial cells. The arrangement of all types of brain cell in three dimensions is crucial. After all, we could have brains that consisted of the right number of billions of both neurons and glial cells that would be totally incapable of thought. Imagine, for example, at the end of one neuron the only contact that was made was with a glial cell. Such an arrangement, if repeated through the entire brain, would be functionless.

The Four-Dimensional Brain

The formation of the large multi-bulged adult brain from the small triple-bulged embryonic one can be thought of as a challenge in the molding of gross shape. But in this respect it's not so very different from the formation of a heart, a lung, or a liver. So we'll concentrate on the cellular circuitry aspect of things, which is what makes the brain unique. As a sort of preamble to this, though, we should look more deeply at an individual nerve cell and the connections, or *synapses,* between two such cells, for these are the units of which the overall circuitry is made.

A typical neuron has a plump cell body in which the nucleus resides and, extending from this, two types of thin, stringy outgrowth. There's usually one long outgrowth, called the axon, and lots of little ones, called dendrites. There's a functional directionality to this structure: the dendrites bring messages into the cell, while the axon sends them out to another cell. In reflex reactions that don't involve the brain, the "circuit" can be as minimal as two or three neurons transmitting a message that starts with an environmental stimulus and ends with a muscle movement. In contrast, the circuits used by Charles Darwin when writing *The Origin of Species* were complex in the extreme.

The signals or messages that flow through nerve circuits are largely electrical. However, at most synapses the flow becomes chemical; then it returns to being electrical at the other side of the synapse concerned.

Because the brain contains many billions of cells, the number of possible circuits is almost infinite. However, given any *particular* pattern of overall circuitry, some cells will not be able to connect, at least directly, with some other cells. And with a different pattern, different connections will be precluded. So we come back to the key problem of how the brain develops, from the perspective of connections among its various neurons.

During embryogenesis, stem cells in the neural tube make nerve cells. These move to the outside of the tube, thus making it thicker. As time goes on, more and more nerve cells are made, and each batch moves out beyond the already-formed outer layer. So, after the elapse of much time, we have a system of layers in which the outermost are the youngest and the innermost the oldest. This layering process is how the cerebral cortex, the key thinking part of our brains, is produced in development.

However, this is only part of the story—the part about how nerve cells are made and how they move to their final positions. The other part of the story of brain development is about the formation of junctions between nerve cells, and hence the formation of circuits. As the dendrites and axons grow outward from nerve cell bodies, they make contact with other nerve cells and can form synapses with them, which will become the basis for nerve impulse propagation. But how do they know which direction to grow in, and how far to grow, before deciding to stop and make a connection? This problem is most pronounced for axons, which are by far the longer of the two types of projection.

A developing axon has a thickened part at the end called a growth cone. It's a bit like the thickened part at the end of a tentacle of a garden snail. You may sometimes have watched these creatures moving around. If one of their tentacles hits an object, the tentacle usually gets retracted. But shortly afterward it starts to project outward again. This outward projection is a bit like what is happening in a growing axon. But in the case of the snail's tentacle, the outgrowth is controlled by the snail's brain, whereas in the case of an outgrowing axon, the fully formed brain of which it will end up as a part does not yet exist. So how does it know what to do?

This is an area of intensive current research and the answer is only very incompletely known. It seems that axon growth cones in a sense explore their environment as they grow. They react to both "guidepost" cells and guidepost chemicals. For example, on encountering a gradient of a guidepost chemical, they may turn a corner so that they head either upstream or downstream in the gradient. Eventually, after responding to many cues, they will form a synapse with another nerve cell. Synapses are variable in form—sometimes an axon will make contact with the body of another cell, other times with one of the other cell's dendrites. Think of this process going on billions of times in parallel, with both commonalities (inbuilt limitations to the nature of axon growth) and differences (different combinations of cues in each part of the developing brain). Will we ever understand it fully? The jury is still out.

When Does Brain Development Stop?

Like all our other organs, the brain continues to develop following birth. Indeed, from the perspective of stimulus-driven development, this increases dramatically after we are born. The womb, as an internal environment for a floating fetus, is not without stimuli, but these are as nothing compared with the wealth of sensory information that bombards us when we make the switch to an external environment. The rate of brain development in our first couple of years after birth is truly amazing. One estimate puts the number of brain cells formed *per minute* during this period as more than 100,000 and the number of synapses formed in that same time in the millions.

After infancy, the rate of formation of new brain cells and the connections between them slows down. But in a way that's not saying much since the initial rate of their formation is so high. And although by the time we're adults it has slowed down a lot, recent studies suggest that the capacity for humans and other mammals to form brain cells in adulthood has not fallen to zero, as used to be thought. So one possibility is that brain development doesn't stop (or go into reverse) until quite late in adulthood, especially for people who exercise their brains on a daily basis.

One especially interesting aspect of brain development is how and when it gives rise to the origin of consciousness. Unfortunately, the "how" question is not yet answerable in any satisfactory way. The "when" question is somewhat easier, but its answer is not, of course, a precise age. Rather, it's a range of ages. Whether the earliest origins of consciousness are pre- or postnatal is hard to know. It's even hard to know what the level of consciousness of a two-year-old child is, since no one can remember that far back. Four years of age, yes; three, maybe (some folk claim they can recall events that happened when they were three); two or one, definitely not.

But such an approach raises the question of what the connection is between consciousness and memory—another tricky issue. Thinking my way back through my own life, I'd say my consciousness began by four years of age, and continued to develop from then to adulthood, with growth spurts around the ages of 12 and 17. How does that compare with your perception of the development of your consciousness? I'd guess it's not too different, and yet this "thinking back" is a very subjective approach. Can it be connected with the scientific approaches of psychology and neuroscience? Not very well as yet, but such connections are growing, and this is a hugely exciting field of research.

A Very Organized Matter

The American biologist Scott Gilbert writes, in his blockbuster 2014 textbook *Developmental Biology,* that "the human brain may be the most organized piece of matter in the solar system" (341). And despite the fact that the degree of organization of something is a hard concept to define, I suspect that Gilbert is right. Sometimes intuitive feelings, themselves interesting products of our complex brains, suffice, and definitions can wait. The human brain seems to be more organized, or complex, than any other animal brain. And it seems to be more organized than the system of rings and moons orbiting Saturn. This second comparison is both easier and harder to make than the first one. It's easier because the difference in complexity between a human brain and an orbital system is much greater than that between a human brain and a chimp one. But it's harder because the nature of the two systems and their components is completely different. Anyhow, I'm convinced that our brain is indeed the most complex or organized object in the solar system, whichever other object we choose to compare it with.

Gilbert's choice of the context for these comparisons—our solar system—was a wise one. It seems likely that in other such systems life has evolved, including, in some cases, intelligent life-forms with brains. Some of these may be more complex by far than ours. Indeed, this seems the most probable situation, even though we don't yet have evidence for it. We humans have had a tendency, in our intellectual development, to see ourselves as the pinnacle of life living at the center of the universe. We've known since Copernicus that the second of these things isn't true. Perhaps we'll soon discover that the first isn't true either.

If intelligent aliens exist, they are almost certainly not to be found in the centers of stars or in interstellar space. Rather, they're likely to be found on planets much like our own but a long way away. And, as of the last few years, we finally know of such planets. The story of their discovery is a fascinating one; it's the subject of Chapter 16.

VI

MILESTONES OF DISCOVERY

Chapter Sixteen

Exoplanets and Aliens

Are We Alone?

For centuries, people have contemplated the existence of other worlds, and considered the possibility that some of these may be inhabited by intelligent life-forms. The ancient Greeks thought about this, as they did about most things. The sixteenth-century Italian friar and philosopher Giordano Bruno proposed that the stars were suns and that each might have its own planets, even with living creatures on them. For these and other progressive thoughts, he was tried for heresy by the Catholic Church, found guilty, and burned at the stake—surely one of the most horrific deaths imaginable. His crime: thinking for himself.

Fast-forward to the twentieth century. Others were proposing similar possibilities for alien worlds and life, but luckily in a more enlightened religious and political climate. The American astronomer Frank Drake, whom we met earlier, in 1961 came up with his famous Drake equation to predict the number of extraterrestrial civilizations. In the 1980s there were several claimed discoveries of exoplanets—planets orbiting suns other than our own—but most subsequently proved to be erroneous interpretations of complex data. An example of this was the proposal, in 1991, of an exoplanet orbiting a dead star about 30,000 light-years away. The British astronomer Andrew Lyne, who initially claimed he had discovered this planet, reanalyzed the data and retracted his claim. He thereby demonstrated the difference between a scientist, whose beliefs are subject to alteration by evidence, and the people who tried and burned Giordano Bruno—their beliefs were of an altogether more immutable, and sinister, nature.

The first confirmed exoplanet discovery was a planet orbiting a closer dead star, this time only 1,000 light-years from Earth, by the U.S.-based astronomers Aleksander Wolszczan and Dale Frail. In fact, this pair discovered *two* exoplanets orbiting the star concerned (which goes by the catchy name of PSR B1257+12). If you want to gaze in its general direction, it lies in the constellation of Virgo. The initial discovery was in 1988, but the planets were not regarded as being fully confirmed as such until 1992. And the 1990s turned out to be the decade in which exoplanet discovery got going in earnest, with most of the early discoveries being hot Jupiters.

Hot Jupiters

Our own familiar Jupiter, that stripy gas giant beyond Mars, is a very cold world indeed, with a surface temperature of -100 degrees Centigrade (-150 degrees Fahrenheit) or lower—though note that what constitutes a "surface" on a planet that's largely composed of gas is a moot point. Saturn is even colder. So among the planets of our own system, we tend to connect big with cold. All the inner planets—Mercury, Venus, Earth, and Mars—are tiny by Jovian standards. But it needn't have been like that. There's no law that says planetary systems must take this form. Indeed, there's probably much historical contingency—or cosmic accident—in determining where the big planets are in any particular system.

Many of the early discoveries were of large planets orbiting close to their suns, so these became known as hot Jupiters. The first to be discovered orbiting a still-alive star, by the Swiss astronomers Michel Mayor and Didier Queloz in 1995, was 51 Pegasi b. At this point we need to learn a little about how exoplanets are named. The system used for 51 Pegasi b is as follows. The initial number and name determine the star and the constellation it's found in. So in this example we have star number 51 in the constellation of Pegasus, the celestial winged horse. The first planet to be discovered orbiting a particular star is called b (because the star itself is thought of as object A in the system); the second planet discovered is c, and so on.

The planet 51 Pegasi b is thought to have a surface temperature of more than 1,000 kelvin. So it's hardly likely to be home to life, at least as we know it. For several years, most of the planets discovered were further hot Jupiters, which, from an astrobiological perspective, was a bit of a disappointment. But our methods of searching for planets detect big ones more easily than small ones, and some of these methods also detect close-in ones with rapid orbits more readily than farther-out ones with longer orbits. The orbital period of 51 Pegasi b is an incredibly rapid four days. Compare this to Earth's 365 days and Jupiter's 11 years. This ultra-brief orbital period is related to the fact that the planet is only about 1/20 the distance from its sun as Earth is from ours.

From the perspective of an astrobiologist, all of the hot Jupiters that turned up later were equally uninteresting. What we'd really love to find are exoplanets that are quite like the Earth, orbiting suns that are a bit like ours, not dead ones as in the case of the very first confirmed exoplanets. The PSR in what I called the catchy name of their host star identifies it as a pulsar—an ultra-dense neutron star that's rotating rapidly and sending out beams of radiation as it does so. Each beam comes in our direction once every rotation—hence we see them as pulses. Pulsars are one type of dead star (another type is a white dwarf), but what we'd prefer to see as a host star to possible life-bearing planets is a main-sequence star, such as our Sun, which is often called a middle-aged star since it's about halfway through its life span. In fact, the host star of 51 Pegasi b *is* such a star, but what use is it to life if the planet orbiting it is so hot that it would instantly incinerate any life-form that was foolish enough to visit it from elsewhere? The combination of an Earth-like planet orbiting a Sun-like star was not discovered until we entered the new millennium. But now we know of several, and mostly because of Kepler.

A Tale of Two Keplers

Johannes Kepler was a German astronomer born in 1571, some 28 years after Copernicus's publication of his idea of a heliocentric solar system. And it can be argued that Kepler and Galileo were the two most important figures in confirming and refining Copernicus's Sun-centered arrangement. Galileo is of course well known for many things, including his discovery of four moons orbiting Jupiter in 1610. Kepler is much less well known, and yet his laws of planetary motion—three of them—are crucial to describing the movements of the planets around the Sun.

Kepler's first law, published in 1609, states that planetary orbits are not perfect circles, but rather distorted ones—ellipses. This is now recognized as generally true, though the magnitude of the departure from a perfect circle differs from one planet to another. In most cases the distortion is slight, as we saw for the Earth in Chapter 13. Kepler's second and third laws are more technical; they relate to quantitative aspects of elliptical orbits. We don't need to examine these here, but it's worth a brief mention that his third law describes the relationship between the size of a planet's orbit and the orbital period, which has been important in recent exoplanet research.

There's a noble tradition of naming spacecraft after famous astronomers. And where we're going now is to that other Kepler—the

Kepler space telescope (KST). But first, just a couple of other cases of such naming. Most famously, there's NASA's Hubble space telescope, named after the twentieth-century American astronomer Edwin Hubble. Less well known but just as interesting is the European Space Agency's Huygens probe, named after the seventeenth-century Dutch astronomer Christiaan Huygens, who discovered Saturn's largest moon, Titan. Right now, this tiny craft is probably sitting in the exact same spot where it landed in 2005 on the surface of Titan, a location that gives it one of its main claims to fame. The Huygens probe still holds the record for the most distant landing from Earth of any human-made object.

The Kepler space telescope is not as well known outside of astronomical circles as its Hubble counterpart. And yet I think it should be, because the KST has probably contributed more to our discovery of exoplanets than any other device. At the time of writing, it has been responsible for the discovery of more than 2,000 confirmed exoplanets, over half of the current total. And it has been responsible for the discovery of several Earth-like planets that may yet turn out to be the homes of alien life.

We noted the planet called Kepler 186f in Chapter 1. Notice that the system of naming is a bit different from that used for 51 Pegasi b. This time, the word in the planet's name refers to the telescope that discovered it, not the constellation in which it was found; however, the letter has the same significance in the two systems of naming. So, f means that the planet was the fifth to be discovered orbiting the star Kepler 186, which is in the constellation of Cygnus the swan.

As of early 2017, the Kepler "hall of fame" of particularly Earth-like exoplanets has increased to about 20. And the most Earth-like of all is no longer Kepler 186f but now arguably Kepler 438b, though this orbits a flare star, making it subject to periodic blasts of intense radiation. Kepler 452b is also especially interesting, as it orbits a particularly Sun-like star. However, it's necessary to appreciate that the KST has not actually *seen* any of these planets. How can we be confident that they exist at all, given the various early claims of exoplanets that turned out to be illusory?

Seeing Is Believing

This question raises the related one of how the KST actually works. So let's now examine this matter and persuade ourselves that its exoplanet discoveries are real. Kepler's mode of planet hunting is ingenious. It looks for little dips in the brightness of a star, which are caused by the transit of an orbiting planet across the "front" of that star from our Earth-based viewpoint.

To understand how this works in a bit more detail, we need to know where the telescope is looking, how far it can see, and what its resolving capability is. First, the field of view. This is an area of sky centered roughly halfway between the constellations of Cygnus the swan and Lyra (which looks a bit like a lyre, or small harp), and overlapping both. The total field of view is less than 1 percent of the whole sky, so the number of exoplanets the telescope can find will be a tiny fraction of those that actually exist in our Milky Way galaxy. The depth of view is also rather small—the KST can only look a short distance through the Milky Way, which reduces the number of stars around which it can search for planets. This number turns out to be about 100,000, less than a millionth of the number of stars in the galaxy.

Kepler's resolving capability is limited by two main things. First, it can't detect tiny planets orbiting huge stars because the proportionate diminution of light reaching it as a result of planet transit is negligible. Second, it can only see larger planets when the system concerned is edge-on. If Kepler is seeing a particular system face-on then there won't, from its perspective, be any transits at all.

For edge-on systems, a single dip in light coming from a star could be due to several causes, so we don't immediately yell eureka when we find such a dip. But if it recurs after, say, six months, we get excited; and if it recurs after another six months, we get very excited indeed. By far the most likely explanation for these observations is that a planet is orbiting the star concerned and that its orbital period is six months—half that of the Earth around the Sun. Nevertheless, dips in the amount of light constitute a very indirect form of evidence. We conclude there's a planet there not because we can see it but because we can see a periodic decrease in the light from the star it's orbiting. This sort of evidence is quite compelling when we've observed multiple orbits of the same planet, but it's not quite the same as *seeing* the planet concerned.

Have any exoplanets been directly observed? Several years ago the answer to this question would have been no, but now it's an emphatic yes. One example is the planet called Fomalhaut b, discovered in 2008 from Hubble images. This planet is very large (bigger than Jupiter) and very far from its sun (more than 100 times the Earth's distance from our Sun), and this system is a mere 25 light-years away, much closer than Kepler 438b, which is nearly 500 light-years away. None of these things are surprising, because big planets orbiting far out from close stars are the easiest for us to see. The significance of such direct observations of exoplanets lies in the "seeing is believing" adage, not in their suitability for life, which is generally very low. So let's now turn to more promising planets for life.

The Goldilocks Zone

When Goldilocks sampled the porridge in the three bears' bowls, she famously found that in one case it was too hot and in another case too cold, but in the third bowl the porridge was "just right." This classification into hot, warm, and cold has been incorporated into the name of a zone around a star in which the surface temperature of a planet would be just right for life: the Goldilocks zone. Or, for those who prefer less literary names, the habitable zone. In our own solar system, the only planet that is squarely within the habitable zone is Earth—though Venus and Mars aren't too far away. Also, complexities can arise because some far-out moons, such as Jupiter's Europa, may be heated internally and thus be warmer than would otherwise be expected for a body so remote from the Sun. But what about other planetary systems? Are there any exoplanets with temperatures of which the cosmic Goldilocks would approve? Well, yes, several have been discovered to date. But there are other criteria for life that we need to take account of as well as temperature.

Let's think about the Earth for a moment, since it's still the only planet we know to host life. It satisfies the Goldilocks criterion, which means that it can have plenty of liquid water on its surface. But it also has a breathable atmosphere, in contrast to Venus, where in an atmosphere largely consisting of carbon dioxide there are clouds and probably rain that consist mainly of sulfuric acid. So the right atmosphere is important too. On Earth the atmosphere has changed a lot over time, largely due to plants, but even the infant Earth had an atmosphere that was suitable for the primitive creatures that inhabited our planet at that time—our distant unicellular ancestors.

Moderate temperatures and a hospitable atmosphere are two of the major criteria for raising the likelihood that a planet might have life, but they're not the only ones. Gravity and pressure are also crucial. The atmospheric pressure on the surface of Venus is about 100 times that on Earth. If we flew there, landed, and exited our craft, we might be crushed before we were poisoned or asphyxiated. And if we flew to an enormous rocky exoplanet with a huge gravitational field, we might collapse into quasi-discs no higher than our ankles.

There are other habitability factors to consider beyond the four we've looked at so far. One is the planet's orbital period. Even if the planet is within the habitable zone, suppose that its orbit takes a week or a decade instead of a year—would life then be less likely? And its rotation on its own axis may also be important, since a day could last for only an hour or for more than a year. Again, it seems likely that the prospects for life would be influenced by these things.

Right now, we know of quite a few planets—about 20 at the time of writing—where the combination of conditions is such that life *should* be possible. Yet despite this we haven't discovered alien life anywhere, not even indirectly via radio transmissions. The SETI endeavor to detect such transmissions has so far drawn a blank. It remains to be seen whether recent extensions of this project, notably Breakthrough Listen—an initiative by Russian physicist and billionaire Yuri Milner—will change that. So while the foregoing parts of this chapter have been firmly based on evidence, the parts that follow are necessarily more speculative.

Alien Evolution

There's a famous, but incorrect, quote from Mr. Spock to Captain Kirk in *Star Trek,* which goes something like this: "It's life, Jim, but not as we know it." Even though what Spock actually said was a bit different and the quote got doctored subsequently, as so many do, it's a useful reminder to us that alien life may not be at all similar to life here on Earth. Remembering this is a good starting point for thinking about the possibility of life elsewhere, as an antidote to our natural human arrogance that all too often affects our search for the truth—such as in early notions of an Earth-centered universe.

That said, looking for life as we *do* know it is probably a more sensible way to start the endeavor than any other, because at least we have evidence that our type of life exists somewhere (here), whereas we don't yet have any evidence that other types of life exist anywhere. But at this point the question arises of what is meant by life that is "not as we know it." This could refer to mammals with six legs or insects with four. At a more fundamental level, it could mean life-forms that do not use DNA as their genetic material. More fundamentally different still would be life-forms that are gaseous rather than solid, such as Abel in the book *Evolving the Alien* by Jack Cohen and Ian Stewart, and those that are not "organic" in that they're based not on carbon compounds of any sort, but rather on compounds in which the central element is, say, silicon.

The insistence that life anywhere in the universe must be carbon-based has become known as carbon chauvinism, with obvious negative connotations. Inasmuch as an awareness of this type of chauvinism helps us to remember that carbon is not the only possible basis for life, that's a good thing. But too much negativity of this kind is unhelpful because, as noted above, looking for carbon-based life is the best way for us to start our search.

As an evolutionary biologist, here's what I suspect alien life, and alien evolution, will turn out to be like when we finally discover it. But remember, this is just a personal view unsupported by any evidence; in other words, it's just a hypothesis. However, like all good hypotheses, it's ultimately testable. Here it is:

Somewhere in our galaxy there is a star very like the Sun. It is orbited by several planets, one of which is very like the Earth. Its surface has many areas of liquid water—oceans. In one of these, or at its edge, large carbon-based molecules stick together and form protocells. Darwinian selection acts on the coherence and replicability of these, and they evolve to become membrane-bounded entities that we can call cells. Eventually, some groups of these cells stick together to form primitive multicellular creatures. Some of these are able to convert solar energy into its biological counterpart—these could be thought of as extraterrestrial algae. Others make their living by eating the algae—these are extraterrestrial herbivores. After the elapse of several billion years, there are diverse large, complex multicellular creatures, at least one of which becomes self-aware. It develops a society, a scientific understanding of things, and a technology that ultimately involves the production of telescopes and spacecraft. It explores its own planetary system, sends radio transmissions into space, and scours the heavens looking for exoplanets (Earth would be one from their point of view) and extraterrestrial life (e.g., us).

These intelligent, self-aware aliens might look familiar or bizarre. They might be eerily humanoid, like the hypothetical Andromedan scientists whom we met in the first chapter. Or they might be distinctly reptilian, like many of the aliens of science fiction. Alternatively, they might be octopus-like creatures whose civilizations began under the oceans. And finally, they might not look like anything belonging to our terrestrial fauna or to the imaginations of sci-fi enthusiasts, and might take a form that we would have difficulty in recognizing as a *life*-form at all.

The Closest Aliens to Earth

Time will tell whether these speculations, or hypotheses, are anywhere near to the truth. But for now there's one final issue for us to grapple with before proceeding to Chapter 17: the question of how far away from us the closest intelligent aliens are likely to be. They're probably closer than Andromeda (the realm of the *far*), but they're almost certainly not on Mars (the realm of the *close*). So we should be looking in the realm of the *middling*—our own galaxy, the Milky Way.

The trouble is that the "middling" scale of our galaxy is vast, and aliens might exist anywhere within it. The Milky Way is about 100,000 light-years in diameter. But we don't need to traverse more than a tiny fraction of that distance to find Earth-like exoplanets. In fact, the closest star system to us, the triple-star Alpha Centauri system, which is a mere four light-years away, was recently discovered to have an Earth-like exoplanet. This has been provisionally named Proxima b (after its sun, Proxima Centauri). The Earth-like exoplanets mentioned earlier, Kepler 186f and 438b, are both roughly

500 light-years away. The most distant exoplanet that has been discovered at the time of writing is about 20,000 light-years from us. In contrast, any Andromedan aliens are more than *2 million* light-years from Earth.

Given that the Kepler space telescope has discovered more than 2,000 confirmed exoplanets, plus a few thousand more probables, within a relatively small section of the Milky Way, in what might be called our immediate vicinity, my guess about the location of the closest intelligent alien life is this: I suspect that it lives within a few tens of light-years from us, or at the very most a few hundreds. Perhaps the lack of radio signals suggests that the latter of these two ranges is more likely. We'll return to this issue of alien radio messages in a later chapter. But next on our agenda is a closer look at the origin and nature of our theory of evolution—a theory that was developed to account for life on Earth but that may turn out to be equally applicable elsewhere. I doubt if biology is a "one-planet science," as some astronomers used to disparagingly claim, and I doubt if Darwin's theory of evolution by natural selection is only applicable here on Earth.

Chapter Seventeen

From Darwin to Darwinism

Five Key Players

There's a great literary leviathan, often referred to as the "Darwin industry." This is the vast collection of historical works on Darwin himself and on the numerous people, both scientists and others, with whom Darwin interacted. Some of the books in this genre are voluminous, and their level of historical scholarship is formidable. Here we'll take an approach that's more appropriate for a popular science book in general, and especially for one with such a broad range of topics to cover. What follows is a short and selective—but nevertheless I believe accurate—account of some of the goings-on in Victorian scientific circles in the mid-1800s that led to the eventual acceptance of the theory of evolution.

We'll restrict ourselves to five players, of which Darwin is of course one. The others might be thought of, in Western movie parlance, as two good guys and two bad guys. The former were "Darwin's bulldog," Thomas Henry Huxley, and Darwin's co-discoverer of the idea of natural selection, Alfred Russel Wallace. The bad guys were a biologist, Richard Owen, and a bishop, "Soapy Sam" Wilberforce. As in other contexts, the good guys are not all good and the bad guys are not all bad, but these movie descriptors are nevertheless rather appropriate in this case. And the main distinguishing features between the two types of player in this hugely important intellectual context were, in my view, not intelligence or diligence but rather honesty and humility—especially the humility to accept that one's own view may turn out to be wrong.

Although the story really starts quite early in Darwin's life, on his heroic five-year journey around the world on the ship HMS *Beagle* in the 1830s, we'll start in the year 1843. This was the year in which one of the bad guys, Owen, published an important book—a collection of 24 lectures given at the Royal College of Surgeons in London earlier the same year.

Different Similarities

Since long before the nineteenth century, those who studied the comparative anatomy of different kinds of animal had recognized the existence of two different types of similarity—those that were somehow deep and meaningful and those that, in contrast, were merely superficial. And it wasn't the case that these were just the opposite ends of a continuous spectrum; they were two distinct types of similarity with no intermediates. One of Richard Owen's claims to fame was that he invented names for them, names that stuck: homology (deep, in some sense) and analogy (superficial). We still use these terms today.

But what do they mean? Owen explains them in the glossary of his 1843 book, under the related terms *homologue* and *analogue*:

Homologue: The same organ in different animals under every variety of form and function.

Analogue: A part or organ in one animal which has the same function as another part or organ in a different animal.

On the surface, these terms don't sound very different from each other, and we could be excused for asking what on earth Owen meant by his dichotomy. But, as ever, things in science become clearer with examples, so let's look at one example of each.

First, homologues. Our human arms are homologues to the front legs of mice and elephants. Or, to put it another way, these structures are all homologous with each other. Today we understand this to mean that the reason all of these structures are built on the same pattern of bones is that we inherited them from a common ancestral mammal whose front legs were built in this way. Thus homology is deeply meaningful because it reflects common ancestry.

Next, analogues. Our human arms are very dexterous and can perform lots of tasks using tools, whether the stone axe-heads of the past or the screwdrivers and keyboards of today. Likewise, the tentacles of octopuses are very dexterous, enabling them to do many things that their snail cousins can't do. And, like our arms, octopus tentacles are at the anterior end of the body. However, the similarity in function is not underlain by a similarity in structure, and the reason for this is that our arms and the "arms" of an octopus were separate evolutionary inventions. Another way of putting this is that the last common ancestor of humans and octopuses—and there really was such a creature—did not have any arms at all.

Owen himself used this particular example, although he did not think of it in evolutionary terms. Here's what he said in the closing chapter of his 1843 book: "The sole locomotive organs [tentacles] in the ordinary Octopods . . . have no true homology with the locomotive members [arms and legs] of the Vertebrata, but are analogous to them, inasmuch as they relate to the locomotive and prehensile faculties of the animal."

Notice the use of "true" but the lack of any evolutionary interpretation. The latter is unsurprising, perhaps, because this was written more than a decade before the publication of *The Origin of Species;*

and even after *Origin* appeared, Owen never accepted Darwinian evolution by natural selection. But in that case, what did he mean by "true"? And is it then the case that mere analogous similarities are somehow "false"? For creationist anatomists, homology somehow reflected the divine plan, whereas analogy represented distractions from it. And yet by the 1840s some biologists were thinking evolutionary thoughts, albeit mostly not based on natural selection, so to them the interpretation of homologues and analogues would have been different.

One such person was Thomas Henry Huxley, a quotation from whom I used at the start of this book. During the early 1850s, for example, Huxley was exchanging correspondence with a fellow expert on molluscs, Albany Hancock, on the issue of which nerves in one group of molluscs were homologous to which nerves in other groups: for example, squids versus snails. These and many other biologists of the time were busy trying to identify similarities that were homologous rather than merely analogous, though without a clear understanding of what Owen's terms really meant. But the answer was soon to emerge.

The Book That Stole the Show

When Darwin published *The Origin of Species* in 1859, it sold out immediately and had to be reprinted, eventually running to many editions spread over a period of several years. Not everyone who read it believed its main thesis, of course, but many did. And others came round to Darwin's way of thinking later. Eventually nearly all biologists recognized its veracity. The success of Darwin's magnum opus is thought to be one of the reasons that Owen came to dislike Darwin—thus making Owen one of the bad guys in my version of the story. Owen was very possessive of the whole field of the comparative anatomy of animals and didn't take too kindly to being scooped in his own field by a younger colleague.

One of the key public gatherings to debate the new evolutionary theory in the months following publication of *Origin* was a debate held in Oxford in June 1860, which involved many participants—but the two who are most remembered for their combative exchange are T. H. Huxley and Bishop Wilberforce. Ironically, Huxley was initially not planning to attend the debate, on Saturday, June 30, because he had plans to travel elsewhere. But a chance encounter on the day before the debate with another evolutionist, who was concerned that the bishop would wipe the floor with any opponents, led Huxley to change his mind and remain in Oxford a day longer to attend the debate. The rest, as they say, is history.

Regrettably, there exists no authoritative written account of what was said during the debate. There are many supposed quotes, but they're not to be trusted. So let's be content with the gist of the most famous part of the encounter. The bishop asked Huxley if he was descended from a monkey on his mother's side or his father's. Huxley replied that he would rather have a monkey as a relative than a man who used his eloquence to belittle honest seekers after the truth.

Various letters subsequently written by people who had been present at the debate refer to ladies fainting and men leaping out of their seats, though these may be exaggerated accounts of the response of the audience as a whole. But the important outcome of the debate was that Huxley was perceived to be the winner, and this was one of many factors that led to more people coming to accept Darwin's theory of evolution.

When Is a Scientific Theory Accepted?

Surveys show that even now, in the twenty-first century, the majority of Americans apparently don't believe in evolution. This situation, in the country that leads the world in many areas of science and technology, including space exploration, is both incredible and appalling to a European such as me. But even here on the Fair

Continent, where evolutionists have the upper hand, there remain significant numbers of people who think Darwin was wrong and who believe the biblical account of creation. These facts raise the question of how we decide if a scientific theory has become accepted.

Actually, the opinions of laypeople are of little consequence in relation to this issue. Science is not a democracy, though it often thrives best under political systems that are themselves democratic. Most Americans who have taken part in surveys about evolution and creation have never read an evolutionary biology book in their lives—so how can they make an informed comment? This is a bit like me being asked what I think of post-modernism, when it's a phrase that I've heard bandied about but I am clueless as to what, if anything, it really means.

So, the question of whether a scientific theory has become accepted hinges on what scientists think, and in particular whether the theory ends up being accepted by the overwhelming majority of those in the field concerned and in fields immediately adjacent to that one. Darwinian evolution theory is now endorsed by such a majority; it can therefore be said to be accepted. But there are two caveats to this apparently simple conclusion. First, "overwhelming majority" is not the same as "everyone." And second, acceptance of a scientific theory is only ever provisional because scientists are prepared to change their minds if new evidence comes along, whereas many religious folk are not. Although the vast majority of physical scientists believed that Newton's laws of motion were universally true, they had to admit that this was not the case after digesting Einstein's theory of relativity.

The Essence of Darwinism

Wait a minute, though. I've said that Darwin's theory is now accepted, but what exactly was (and is) his theory? The answer to this question is not as simple as it might initially seem. In a way, Darwin had three theories. First, he espoused the theory that evolution—his "descent with modification"—had actually happened. He was by no means the first person to propose this; his forerunners in this respect were many, and included his grandfather Erasmus. Second, he proposed that natural selection—"survival of the fittest," as some came to call it later—was an effective evolutionary agent. Third, and most important, he connected these two things together and claimed that "natural selection has been the main but not exclusive means of modification."

This last proposal is the real essence of Darwinism. What Darwin meant by it was that if we look at any evolutionary transition, whether from protohuman to human or from snail-like ancestor to octopus, the main mechanism causing the changes that happened in the lineages concerned was natural selection and not something else—whatever that something might be.

Huxley was convinced by this "essence of Darwinism"; and, partly as a result of Huxley's own endeavors, so too were an increasing number of Victorian naturalists and scientists. A century after the publication of *On the Origin of Species,* biologists had come to a consensus that Darwin was right. This mid-twentieth-century consensus is generally referred to as "the modern synthesis."

Of course, from our twenty-first-century perspective, the 1950s and 1960s are hardly modern. So what is the situation today, half a century later? Do we still believe that natural selection is the main means of evolutionary modification? Well, yes, though with a few ifs and buts, and with a healthy debate about exactly which is the most significant of these qualifications. But one thing that is not at issue among professional biologists today is whether Darwinian natural selection applies to all an organism's characteristics. Natural selection has modified human arms and hands; it has also modified human legs and feet. To claim that it has affected one but not the other would seem to be folly. And yet there's a not-too-different claim to which some folk give credence.

Evolution of the Mind

I'm referring here to the claim that the whole of the living world, with its millions of species of animals, plants, and microbes, has evolved by Darwinian selection, but humans—either in general or just our minds—have not.

Now, this belief seems to me to represent an appalling piece of human arrogance. And yet I can understand why some people adopt it—even the co-devisor of the theory of natural selection, Alfred Russel Wallace. He agreed with Darwin about the evolution of the human *body,* but not about the human mind, with its capability for abstract thought, moral judgment, mathematical and musical ability, and love. Wallace believed that the human mind belonged to another world—the spiritual one—and that its origin and evolution could absolutely not be explained in purely physical terms. He makes this very clear in the final chapter of his 1889 book, which, ironically, is entitled *Darwinism.*

The main reason that people take this stance is that to do otherwise—to accept a Darwinian explanation of our minds—would seem to deal a fatal blow to any hope of there being life after death. Let's face it, the idea of nonexistence is not very appealing to most of us. And it gets progressively less appealing with age, since it feels like a much more imminent fate than it did when we were teenagers and our earthly life seemed to project so far into the future that we could barely envisage its end. But fear of possible oblivion is not a good basis for decisions about our understanding of the universe in which we live, including our understanding of the nature of human life. So I'm going to proceed on the basis that Darwin was right and Wallace was wrong about the evolution of the human mind. Strangely, though, this stance does not lead me to abandon agnosticism for atheism. Once again, some wise words from T. H. Huxley—the inventor of the term *agnosticism*—are worth quoting: "I am too much of a sceptic to deny the possibility of anything" (L. Huxley 1900, 2:127).

Applicability to Aliens

The concluding section of the previous chapter was entitled "Alien Evolution." In it, I gave a hypothetical scenario for the evolution of intelligent extraterrestrials on a planet at an undisclosed location in the Milky Way. This scenario incorporated Darwinian selection. Let's now pause to ask whether it is reasonable to extrapolate the basis for life on Earth to life elsewhere in our galaxy, or, indeed, in the universe as a whole.

As mentioned earlier, some astronomers have referred disparagingly to biology as a one-planet science. The rationale for this (usually jocular) view is that while the laws of physics seem to apply everywhere, the laws of biology may not, because so far we only know of a single planet on which we can test them. If there are other planets with life, which there almost certainly are, might that life not be so different from life here on Earth that none of our terrestrial biological laws apply to it? My answer to this question is yes. But if I'm asked the subtly different question of whether our terrestrial biological laws would *probably* apply to alien life, my answer would also be yes, though with a few qualifications.

There aren't many (or any?) true laws in biology in the sense of general principles that are both quantifiable and universally applicable within the living world that we know. But there are some theories that come close. Evolution by Darwinian selection is one of them. The cell theory, which states that all organisms are made of one or more cells, is another. And Mendel's theory of inheritance (actually two interconnected "laws") is a third.

Which, if any, of these would we expect to apply to life on other planets? My guess is that they are all likely to do so, but if I had to rank them I'd say that Darwinian selection is the most likely; I'd rank the cell theory in second place, and Mendelian inheritance third. Of course, many biologists might rank them differently, but few would consider them all to be inapplicable to alien life. The quasi-laws of biology may yet turn out to be applicable on distant planets both within our own galaxy and in others, be they neighbors like Andromeda or so far from us that not even our most powerful telescopes can see them.

The thing about natural selection that makes it so likely to apply to alien life is that the features required for it to happen are very few and very probable. In fact, there are really just three of them, as noted earlier: variation, reproduction, and inheritance. If a group of entities has these three features—be such entities bacteria, sponges, humans, or aliens—then natural selection is not just possible, it's inevitable. If parents give rise to variable offspring that on average resemble their parents a little more than do the offspring of other parents, then Darwinian selection will happen, and if the planet concerned is stable enough for selection to continue for billions of years, the range of its possible products, while not limitless, is very wide indeed. So perhaps we should expect that on every exoplanet with life, evolution by Darwinian selection will occur. The creatures it will lead to will not be the same as here on Earth, though we're still not sure whether to expect them to be very different or only a little different.

Although we live in the space age, while Darwin, Huxley, and their colleagues lived in an era before the first planes flew, I'd bet that these nineteenth-century "good guys" speculated about the possibility of extraterrestrial life and the question of whether its evolution was powered by natural selection. Yet there were two important things that they didn't know about life on Earth: the mechanism of *inheritance* from parents to offspring and the mechanism of *development* from egg to adult.

Missing Mechanisms

The missing mechanism for inheritance was not long in coming, though when it came it was long ignored. We briefly met Gregor Mendel, the monk-scientist, in Chapter 6. Mendel's experiments on inheritance in pea plants were the foundation of genetics. Their results were published in 1866, but his paper was largely overlooked, and only "rediscovered" in 1900. Mendel conducted the experiments on which his laws of inheritance were based in the late 1850s and early 1860s. But they were carried out in a monastery garden, not in a university department. And he published them in a relatively obscure journal.

Both Huxley and Darwin were very conscious of the need for a better understanding of inheritance than was available in the 1850s. Huxley wrote to his friend the botanist Joseph Dalton Hooker in 1861: "Why does not somebody go to work experimentally, and get at the law of variation for some one species of plant?" (L. Huxley 1900, 1:227). Unknown to both of them, Mendel was already doing what Huxley was suggesting. Such lack of communication between scientists interested in similar problems is less likely in today's electronic age, but it does still sometimes happen.

The mechanism of inheritance that we owe to Mendel was later confirmed and extended to animals as well as to other plants. And it was given a molecular underpinning by the work of James Watson and Francis Crick, co-discoverers of the structure of DNA, in 1953. But in the 1950s the mechanism of *development* was still largely obscure. We'll now look at some of the important milestones in arriving at our much improved, but still far from complete, understanding of how development works.

Chapter Eighteen

Analyzing the Embryo

The Biggest Unsolved Problem?

When I was a student, the question of how a fertilized egg became an adult, via a series of developmental stages, was described as the biggest unsolved problem in biology. After all, by then (the early 1970s) we understood a lot about how evolution and inheritance worked, and we knew that genes were important players in both of these processes. We also knew that they must be important players in development, both embryonic and post-embryonic, but we really didn't know much about what their roles were.

Today, a lot has changed. We understand that some genes have no, or negligible, roles in the developmental process. These are referred to as *housekeeping* genes, because they produce things such as routine metabolic enzymes that keep our bodies trucking along from day to day. More important, we understand that other genes—*developmental* ones—help to steer the embryo's course from fertilization to birth (or hatching) and ultimately to adulthood. One of these genes—the amusingly named *sonic hedgehog*—was considered briefly in Chapter 3. Here we'll take a look at some of the key milestones on this exciting journey of discovery, from the early days, when the mechanism of development was inscrutable, to the present day, when we understand quite a lot about it. So we'll start when there was no such word as *gene*, and we'll end in the era of the genome.

What's Inside a Sperm?

In the early days of microscopy, some scientists who examined sperm cells at what seemed then to be spectacularly high magnification thought they could see things inside. Although these folk weren't all of one mind about *what* things, a consensus grew among some of them that they could see humans, perfectly formed but very small. This school of thought came to be called preformationism.

However, there were two competing preformationist theories. The one described above is *spermism*, while an alternative was *ovism*. In both, preformed humans were reckoned to exist, but according to one of them these were inside the sperm, while according to the other they were inside the egg. From our present-day perspective, the latter would seem to make (marginally) more sense, because the egg is huge compared to the sperm. Thus, given that having a perfectly formed human in miniature would be tricky, it would be at least a little less so in a big cell than in a small one.

We now know that both spermists and ovists were wrong; hence the entire preformationist school of thought was wrong. There are indeed structures present inside egg and sperm cells, but they're emphatically not little people. Rather, they are things like nuclei, with their constituent chromosomes, and, especially in the case of eggs, a whole lot of other structures too. We can still see pictures of the little humans, or *homunculi*, that some early microscopists thought they could see inside sperm—these images are only a click away—but we now know that they belong in the realm of either history or fiction, depending on your point of view, and that they have no place in the realm of modern science.

Had preformationism turned out to be right, the developmental process that we would have tried to understand would have been simpler and would have consisted largely of growth. However, since it turned out to be wrong, the developmental process that we really have to understand is more complex by far, and consists of many things, including some that we've already looked at in earlier chapters. But one that we haven't yet looked at is the role of a little bit of embryonic tissue called the *organizer*.

Discovering the Organizer

Experiments with amphibians have played an important part in many advances in understanding embryogenesis. We noted earlier that the fist cloned animal was not Dolly the sheep but rather an amphibian to which I gave the name Freddy the frog—a fictional name for a real creature produced by the research group of British embryologist John Gurdon. This work was done in the mid-twentieth century.

But a few decades earlier, in the period from the early years of the new century up to the 1920s and beyond, another research group was at work on amphibian development, led by the German embryologist Hans Spemann, a key member of his group being his doctoral student Hilde Mangold. These researchers were investigating the stage of embryogenesis that we call gastrulation. You might remember this from Chapter 3, though we've covered a lot of territory since then, so here's a brief reminder. The embryo, whether human or amphibian, passes from an early stage that looks like a raspberry or mulberry through a stage called a gastrula and on to a stage that I referred to earlier as a little sausage, in which the central nervous system is forming. One of the many key changes that occur at the gastrula stage is the transition from a spherical shape to a sausage shape—that is, the embryo elongates and develops its main body axes during this phase of its development. So important is this stage that the embryologist Lewis Wolpert is said to have remarked, with his tongue no doubt at least partly in his cheek, that it is "not birth, marriage, or death, but gastrulation, which is truly the most important time in your life."

One of the key players in making gastrulation happen is a little patch of tissue, initially consisting of only about 10 cells, found in a very specific location in the early amphibian embryo. This is the *organizer*—sometimes called the Spemann organizer in honor of its discoverer, who won the Nobel Prize for this research. The organization that it carries out, and the mechanism by which that organization works, are complex and we won't go into them in detail. But one particular type of experiment reveals particularly clearly how important this little bit of tissue is.

Hilde Mangold took a normal newt embryo at a stage prior to gastrulation and transplanted into it the organizer from another newt embryo, grafting it almost exactly opposite the site of the recipient embryo's own organizer. The result: a post-gastrulation embryo that was double-dorsal (in other words, had two backs); if such an embryo develops further it looks like a pair of Siamese-twin newts, conjoined stomach to stomach. Somehow the new organizer had instructed the cells around it to do something different from what they would normally have done. This influence of some cells on the developmental fate of others is called embryonic induction. And we now have lots more examples of it, and in a much wider range of animals, than we did in the early twentieth century. It's one of the most important mechanisms underlying development in general.

The French Flag

One way that a process of induction can work is through a mobile substance that is secreted by the organizing region, travels to the responding region, and instructs the cells there to do certain things. The mobile agent is called a morphogen—literally, "generator of form." It could travel in various ways depending on its size: a little molecule might passively diffuse through tissue, while a big one would probably need some form of active transport. But the general idea is the same, that one bit of an embryo can influence another by acting as the source of a mobile agent.

This notion was given a very vivid mental image in the French flag model, devised by Lewis Wolpert, whom we just met. Here's what the model involves. Consider a rectangular sheet of tissue consisting of many cells. Now imagine a row of cells along one of the short sides of the rectangle; these cells are specialized to secrete a mobile agent that passes through the whole sheet. Because this agent is made at one end of the sheet but not the other, a concentration gradient will be set up with a high level of the mobile agent near where it is produced and a low level at the other end.

It's an easy matter, then, to turn the sheet into a French flag (or any other tricolor; it could equally well be Irish or Italian). All we need are two threshold levels of the mobile agent to which the cells of the sheet respond. If one side of the sheet is attached to a flagpole, and it's the pole cells that make the morphogen, then the tissue at the pole side of the sheet experiences the highest concentration and turns blue; the tissue in the central band experiences a middling level and remains white; and the tissue at the free side experiences the lowest level and turns red. The only proviso here is that there are two definite thresholds in the level of the morphogen to which cells respond.

The trouble is that, like any model, this one is rather abstract. If mobile substances are traversing the embryo, or parts of it, and causing particular developmental fates, what are they? At the time when Wolpert proposed the French flag model, in the 1960s, we didn't know. But now we know of lots. The sonic hedgehog protein, which we encountered in Chapter 3, is one of them, though it wasn't the first to be discovered. The first persuasive morphogen was a substance called retinoic acid that underlies developmental processes in the wing buds of embryonic birds, among other places. A commentary published in the scientific journal *Nature* in 1987 alongside a technical paper reporting this discovery was entitled "We Have a Morphogen!" (and yes, the exclamation mark was there in the original).

Some morphogens are the direct products of genes; others are not. So the discoveries of morphogens have helped us to understand the roles of some genes in making mobile agents that drive the developmental process. But not all developmental genes work in this way. Other key developmental genes do something different but equally important: they make proteins that move into the nucleus and switch other genes on and off. One important group of such genes is characterized by possession of "boxes."

Boxes in Genes

As we've already seen, each gene in the genome of a human, an amphibian, or any other creature, is a linear sequence of building blocks, and different genes have different sequences. Any one gene in any one animal has a precise sequence through which it can be identified. Each gene's sequence is very long—typically many thousands of blocks. So far, for example in Chapter 14, we've cleverly managed to deal with topics involving gene sequences without asking what the building blocks are. But now we need to give these blocks labels so that we can distinguish one from another.

If we unwind the famous double helix of a gene and look at just one of the two strands that are normally twisted around each other, we find that it exhibits a linear sequence of four chemical units, each repeated many times. The chemical names are usually abbreviated to their initial letters, A, C, G, and T, but for those who prefer to see where the initials come from, the names of these chemical units are adenine, cytosine, guanine, and thymine.

Here's a sequence of 12 units somewhere in the middle of an unspecified human gene:

TGCGATGAGAGA

If we look at the same portion of the same gene in another animal, we'll find some similarities and some differences, the ratio of these depending on how closely related the other animal is to us. Here's what we find in the other animal:

GACTCCGAGAGA

Suppose now that we wish to highlight the stretch of the human gene that is the same as its counterpart gene in the other animal. We could do this in several ways, one of which is to use bold font, as follows:

TGCGAT**GAGAGA**

Or we could use an underline, or even draw a rectangle, or box, around the GAGAGA sequence, which the two genes have in common.

Since the early days of molecular genetics, it's been common to draw boxes around a particular stretch of sequence that we wish to highlight, regardless of the reason we're interested in that particular stretch. It might be to emphasize similarity between different animals, as above; it might be to emphasize similarity between two genes in the same animal; or it might be for some other reason.

Now we get to a particular box that's not just 6 units long but 180. This box shows up in many human genes and indeed in many genes in all animals. It's not always *exactly* the same, as our GAGAGA box was in the above simplified and hypothetical example, but it's similar enough to get boxed. This particular box is called a homeobox. It's found in many, but not all, genes that help to control embryogenesis.

Since the genetic code, like this book, works in triplets, 180 units in a gene correspond to 60 units in the gene's product, the protein. The proteins that have a certain stretch of 60 units within them because they were made by genes with 180-unit homeoboxes all have something in common: they bind to DNA. Such binding is what is required of a protein if it is to be able to switch genes on or off. And that's exactly what these proteins do.

So some developmental genes make mobile proteins such as sonic hedgehog that work by leaving the cells they're made in and traveling to others, where they produce a developmental effect. Other developmental genes make more locally restricted proteins that usually stay within the cell in which they were made, going back into the nucleus from which the message to make them emanated, attaching to other genes there and regulating their activity. One example of such a protein in flies is antennapedia, so called because when the gene producing it goes wrong, legs grow out of the head where the antennae should have been. There's an important general principle here in genetic research: what happens when a gene goes wrong tells us a lot about which developmental processes it normally controls.

Mobile proteins and gene-switching proteins don't act in isolation; rather, they interact. Both take part in the process of developmental signaling that we had a quick look at in Chapter 3. A common sort of interaction is this: a mobile protein leaves cell A and ends up at the surface of cell B. There, a receptor protein protruding from the cell's surface attaches to it and sends a signal into the cell's interior to say that this attachment has taken place. Within the cell, the signal is passed on like a baton in a relay race, and eventually a gene-switching protein in the nucleus of cell B switches on a gene that was previously off. Of course, that causally downstream gene might make a mobile protein that heads off to cell C, and so on. This is a very simplified view of the signaling interactions that happen in any embryo, as those form a complex network rather than a linear series; nevertheless, it captures the essence of the process.

The homeobox was discovered in the early 1980s, and, like many scientific discoveries, it was made in parallel by two research groups, in this case one working in Switzerland, the other in Indiana. Perhaps this serves to remind us of the independent "discovery" of natural selection by Darwin and Wallace in the nineteenth century. Anyhow, the homeobox discovery was one of the main factors that led to a whole new branch of biology—the one that's now nicknamed "evo-devo."

What's Evo-Devo?

In the nineteenth and early twentieth centuries, much interesting scientific work was carried out in the field of comparative embryology. One of the most exciting discoveries was that the early embryos of animals that would end up very different, like mice and humans, were rather similar. But the work was all of a descriptive nature: its results could be summarized in the form of drawings, but these gave no clues about the mechanisms at work behind the various embryonic transformations that were portrayed.

Two of the main players in this field in the nineteenth century were Karl von Baer and Ernst Haeckel, the latter of whom we met briefly in Chapter 9. Although there are English translations of some of Haeckel's works, such as *Anthropogenie,* both his and von Baer's most important books were written in German and to date neither has been translated into English in full. However, moving ahead into the twentieth century, the influential British comparative embryologist Sir Gavin de Beer published a book in 1930 called *Embryology and Evolution* (later editions appeared under the revised title *Embryos and Ancestors*).

De Beer's work was more accessible to the English-speaking world than that of his Germanic predecessors, but it was still in the same mold—that is, descriptive. The third and final edition of *Embryos and Ancestors* appeared in 1958. Although embryological knowledge had greatly expanded in the nine decades between 1866, when Haeckel produced his most important book, and 1958, and genetic knowledge had mushroomed over the same period, in the 1950s we still knew nothing of the causal agents of development discussed above. We were still ignorant of morphogens and developmental genes.

The publication of the discovery of the homeobox in 1984 changed all that and had a major rejuvenating effect on the field. So major, in fact, that the title of the field changed to reflect its enhanced ability to deal with causal as well as descriptive matters: *comparative embryology* gave way to *evolutionary developmental biology*. But the new title, while accurate, was a bit of a mouthful—hence the invention (it's not certain by whom) and persistence of the nickname *evo-devo*.

Evo-devo is my own specialist field and I'd love to say lots about it at this point, but instead I'll say something very brief. I and many other proponents of this newish branch of biology have written whole books on the subject to which you can turn for more detailed information should you wish—for example, Sean Carroll's *Endless Forms Most Beautiful: The New Science of Evo-Devo*. But in-depth coverage would be inappropriate here. Rather, we just need an overview.

Evo-devoists take many different approaches to trying to understand the relationship between the two great creative processes of biology, development and evolution. One way to look at the whole endeavor is as an attempt to integrate the *individual* animal and its development with Darwinian theory, which is centered on the action of natural selection on *populations* of animals. Biologists often distinguish between the origin of variation and the action of selection upon it. The latter is often called "survival of the fittest," though Darwin didn't use this term. Another of Sean Carroll's popular science books is called *The Making of the Fittest*. That title illustrates the connection between Darwinian and evo-devo approaches very well.

But the science of evo-devo links with homology just as much as (some would say even more than) it does with natural selection. And although I portrayed the father of homology, Richard Owen, as one of the "bad guys" in Chapter 17, he wasn't all bad, and his concept of homology remains central in evolutionary theory. In the age of evo-devo, we can ask questions like: is the development of homologous structures, such as our arms and a horse's front legs, underlain by the same developmental genes? The answer, as might be expected, is yes. However, it turns out, rather confusingly, that non-homologous limbs, such as those of humans and flies, are also to some extent built by the same genes. This was unexpected and has opened up a whole new raft of exciting questions about how evolution works, especially concerning the interactions between how it works at genetic, developmental, and population levels. We're only just beginning to find answers to some of these questions; there are enough unresolved issues to keep evo-devo in business for a long time to come.

Future Imperfect

One thing that no branch of evolutionary biology, not even evo-devo, tries to do is to predict the future—except in the very short term. The long-term products of evolution on our planet that may appear over the course of millions of years are impossible to foretell. We might, of course, suggest that where science fact ends, science fiction begins. A common assumption of sci-fi adherents is that humans will develop even bigger brains than we have now. They may well be right, though that's only one possible evolutionary track into our future.

There are some *general features* of evolution, however, that may be predictable on the basis of extrapolating what we see in the past into the future; though nothing as specific as the forms of future humans or other animals. Rather, the sorts of things we can predict are more of a statistical nature. For example, we know that the number of present-day species is a small fraction of all the species that have ever lived. Given this fact, it seems that the most common fate of any species at any point in evolutionary time is extinction. We might even go further and say that the eventual fate of *all* species is extinction, given enough time. As far as we know, there are no species alive today that were alive in the Cambrian oceans of 500 million years ago. If that's any guide to projecting forward in time, then no species alive today will still be around 500 million years into the future. In Chapter 20 we'll look at whether that's likely to be true for humans. But before that we'll take the optimistic view that humans will be unique and will survive for much, much longer—for 10 times 500 million years—so that we witness the end of the world.

VII

ENDINGS AND ENLIGHTENMENT

Chapter Nineteen

The End of the World

Streets of Belfast

Here are three things that we know about the demise of our planet: first, it *will* happen; second, we don't know exactly when; third, unless we're very unlucky, it won't be anytime soon. This last point means that the sandwich-board men that I encountered on the streets of my home city when I was a child were wrong.

The men wearing huge sandwich boards proclaiming "The End of the World Is Nigh" seem to have gone now. If anyone can be spotted wearing sandwich boards, they're probably advertising Luigi's Pizza Palace or some other gastronomic establishment. In between the two uses of sandwich boards, Belfast became notorious for the terrorist atrocities of "the Troubles," an issue we'll have a brief look at in the final chapter.

As a child, I was fascinated by the old-style sandwich-board men. At first, when I heard them shout the message that was on their boards, I wondered if they thought that the end of the world was actually in progress, because the words "now" and "nigh" said in a thick Belfast accent sound more or less the same, and I'd never come across the latter word. After my parents explained to me that *nigh* was in fact an archaic word for "near" (in time), I wondered how near they thought it was. At that age, I didn't have the courage to ask them. They seemed somehow unapproachable; somewhere on their boards, along with the predicted end of the world, was nearly always "Repent."

To be fair to the sandwich-board men, their "world" may not have been our planet but rather our species, in which case they might not be too far wrong. But with regard to the Earth, they're probably off by about 5 billion years.

A Survivable Collision

Over the next 3 billion years, the Earth looks about as safe as anything can be in space. Doubtless there will be quite frequent impacts of large meteorites, like the one that exploded over Chelyabinsk, Russia, in 2013, injuring more than 1,000 people, though remarkably killing no one. Almost equally certain is that huge asteroids like the one that impacted what is now Mexico 65 million years ago and caused widespread extinction of many kinds of animals, including the dinosaurs, will hit us very occasionally. But even impacts of such large objects, measuring several kilometers in diameter, are of little consequence to the survival of the planet itself. Of much more concern to the future of the Earth is the behavior of ultra-large objects, by which I mean objects that make the Earth look small.

In about 4 billion years, an extremely large object—itself a collection of large objects—will pay us a visit; in fact, this is an understatement, because the visit may be quasi-permanent. This object is none other than the Andromeda galaxy, with which the book started. As a reminder, Andromeda is the largest member of our local group of galaxies; its number of constituent stars is even higher than the Milky Way's 400 billion or so. When the two galaxies collide, there will be a combined total of more than a trillion stars in our region of space, with the Sun just one of them. And the combined number of planets will probably be bigger still, several trillions, one of these being our Earth in its new double-galactic context.

Strangely, our descendants, if there are any of them left by that time, are unlikely to be troubled by the collision. Galaxies do indeed have colossal numbers of stars, but the space between stars is much more extensive than the volume of the stars themselves. For example, in our corner of the Milky Way, the distance from the Sun to the nearest other star is a little more than four light-years. The number of solar diameters that would fit into such a distance is many millions. In other words, if we consider a "finger" of space that is very long and thin, stretching from here to the Alpha Centauri system, with a single star, the Sun, at one end of it and a triple star, Alpha Centauri, at the other, this finger contains a total of four stars; all of it, except its two tiny ends, is devoid of both stars and planets. In proportionate terms, this finger of space is way more than 99.99 percent empty. And we have no reason to believe that our little corner of the Milky Way is atypical, so our galaxy as a whole is also more than 99.99 percent devoid of large objects.

The situation in the Andromeda galaxy is the same. So it is just as diffuse an object as the Milky Way. When two objects, each consisting of more than 99.99 percent space, collide with each other, the result will be a deafening silence. Almost no star or planet in one galaxy will collide with any large object in the other galaxy. There may be a few unlucky exceptions, but the chances are that both the Earth and the Sun—indeed, our entire solar system—will survive the collision unscathed.

Both Andromeda and the Milky Way are *spiral* galaxies. When two of these collide, the result will be a single giant galaxy, probably an *elliptical* one. We already know many of these, and we suspect that they often form through collisions between two or more spirals. And, speaking of more, it's possible that the third-largest galaxy in our local group, the Triangulum galaxy, will join the party when we fuse with Andromeda.

If there are still humans on Earth after the collision, they'll see a nighttime panorama that's even more majestic than ours today, because the stars that we can see now will then be supplemented by many others. Our descendants will be inhabitants of Milkomeda, and there will be no way of telling, just from looking at a star, which of the two galaxies it came from. Whereabouts our solar system will be in Milkomeda is uncertain, though some scientists have tried to predict just that. But the thing is that it doesn't really matter. How close a planet is to a sun is crucial for life; how close that sun is to the galactic center is much less so.

A Terminal Swelling

So, 3 billion years hence, the Earth probably remains intact. And by 4 billion years into the future that's still likely to be the case, regardless of the galactic merger. But five is our unlucky number, because approximately 5 billion years hence the Sun will die. And the fate of the Earth is deeply connected with that of the Sun. When one dies, the other will die too. But exactly *how* will they die? We have a reasonably clear idea of the nature of this far-off event—called here a terminal swelling of the Sun.

How a star dies depends on its size, as we noted in Chapter 4. The Sun isn't big enough to end its life in a supernova explosion, but its protracted death will be pretty dramatic all the same. This is where we need to take a closer look at the old stars called red giants, because these reveal the fate of our currently middle-aged Sun in about 5 billion years.

There are plenty of red giants that you can easily see in the nighttime sky, clouds permitting. And although they don't look seriously red, they do indeed look a sort of yellowish-orange, and can be distinguished in this way from the majority of stars, which appear white. One that's easy to find is Arcturus. The best way to locate it is to do what astronomers call *arcing to Arcturus*. The starting point for this is the grouping of stars called the Big Dipper, or the Plough, depending on which country you live in. This grouping is part of the larger constellation known as the Great Bear, most of whose other stars are comparatively faint.

The handle of the Big Dipper isn't straight, it's an arc. And to find Arcturus you follow this arc away from the dipper's bowl, continuing the same approximate degree of curvature, until you see a very bright and definitely yellowish-orange star. This is the red giant Arcturus; it's hard to confuse it with anything else.

Arcturus wasn't always its present color. Nor was it always its current giant size. Earlier in its history it was smaller and whiter, and also, though some folk find this counterintuitive, it had a hotter surface. Stars are a bit like Bunsen burners in their color-coding—bluish white is hotter than orangey red. So red giants are huge, but they're not as hot at their surfaces as younger, whiter stars. Having said that, they're extremely hot in their cores, but we can't see these directly.

What we see in the sky tonight, or any other night, is like a still photo providing a record of one tiny time-slice in the history of any star. A movie would better enable us to see its whole history, but it would take millions or billions of years to make, and most of the footage would be very boring. The working out of the life cycles of stars, based on observations of lots of stars at all different stages of their lives (possible) rather than a single star followed right through its life (impossible), is one of the great triumphs of astronomy.

In about 5 billion years the Sun will reach the red giant stage of its life. This means that it will swell to occupy a much larger volume of space than it does now. The radius of the dying, enlarged Sun will be approximately the same as the distance from the Earth to the Sun today. So instead of being comfortably far out in the habitable zone, we'll be uncomfortably close to, or even engulfed by, our local star. The Earth at this stage will probably be incinerated, though there are various other possibilities, such as the blasting off of its outside layers, leaving only a small planetary core.

The route taken by a star as it leaves the main sequence and evolves into a red giant is not a simple one. An initial expansion is followed by a contraction and then another phase of expansion. In the Sun's case, if the first of these doesn't engulf the Earth, the second probably will. However, which of the two swellings is terminal for the Earth is what you might call an academic point because our planet will be uninhabitable long before the first swelling reaches its peak.

Why should a star undergo huge changes in its volume as it dies? Given that it's dying because it's running out of the fuel, hydrogen, that powered its nuclear reactor through most of its life, why doesn't it just shrink? Well, eventually it will shrink to a white dwarf. But not just yet—not until after the swellings of the red giant phase are over.

The *instability* of the size of a star toward the end of its life is best explained with reference to its *stability* in size right through its main-sequence phase. Our Sun at present stays the same volume from millennium to millennium because the force tending to make it smaller, gravity pulling inward toward its core, is balanced by the force tending to make it bigger, pressure pushing outward. But the outward pressure is complex; three different kinds of pressure may be involved, one of which is called radiation pressure—the pressure of the nuclear furnace.

If the outward pressure reduces, the star contracts, but as it does so it becomes denser, and outward pressure increases again. In this way the star convulses. Related to this is the switch between different kinds of nuclear fusion. When hydrogen is exhausted in the core, it starts to burn in a shell around the core; and later the star gets hot enough in its interior to start fusing helium into carbon. Helium burning can sometimes proceed violently, giving rise to what's called a helium flash.

After all the convulsions, the dying star will shed a huge amount of gas and dust into surrounding space, and the stellar remnant, in other words the collapsed core of the star, will become a white dwarf. The dwarf is of little consequence, but the ejected material—the star-death nebula—is important because it will contribute to the birth of the next generation of stars, as we saw in Chapters 4 and 5. Let's have another quick look at this recycling of stellar material.

Children of the Sun

We saw earlier that dying stars eject material in the form of gas and dust that will contribute to the great molecular clouds from which the next generation of stars will be born. And we also saw that each generation of stars has a higher metallicity than the one before it, due to the manufacture of metals (in astronomers' usage of the word) as stars age and die. But is there a close spatial relationship between the locations of a particular star (e.g., the Sun) and its "children"—that is, the later stars that contain material from it?

The answer is not necessarily. Gas in space doesn't always stay still, even if we knew what *still* means in space. It can be buffeted by various forces, one of the most powerful of which is a supernova explosion. The famous star Betelgeuse, in the constellation of Orion the Hunter, will probably die in a supernova explosion within the next few tens of thousands of years. When this happens, Betelgeuse will appear to humans for at least a month as an object that's brighter than the Moon, even though it's more than 600 light-years away. The shock wave will blast gases and dust, not just from itself but from a large volume of interstellar medium in its vicinity, outward at an incredible speed. They will travel long distances and end up far from the original location of the ex-star that was called Betelgeuse.

After our Sun has died and shed a nebula of gas and dust into space, somewhere in its vicinity a larger star will die soon after, and its supernova explosion will blast our remains a long distance in an as yet unknown direction. The children of the Sun will still be in

Milkomeda, but not at the same spot as the Sun is right now in its middle age, which in any event is not a fixed spot, because even boring middle-aged stars are always on the move within their galaxy.

Immortal Astronauts?

One way or another, it's clear that humanity's days are numbered here on Earth. If we survive for the next 5 billion years, which would make us easily the most long-lived Earth species ever, we won't survive the Sun's death at that point in time. So if humanity is to last indefinitely, our only hope is to colonize other worlds, that is, exoplanets. Each of these will be subject to the same life span constraint as Earth—it will only last as long as its sun does. So it might be necessary for us to planet-hop every few million generations to find a new home.

If the tricky technical problems of such long-distance space travel can be solved, the solution possibly involving warp speed or wormholes, then a nomadic long-term future for humanity might be possible. But there's a big difference between long-term and immortal. Space-roaming humans of the distant future would doubtless be individually mortal, just as we individual humans alive today are; their civilizations would be mortal too, and not just because each exoplanet they were on would eventually die. Current theories of the future evolution of the universe suggest that it will continue to expand and cool, eventually reaching a state that has been called the big freeze. At this stage there are predicted to be no stars left, only black holes, and even these may eventually dissipate.

So, as once before in the book, we end with a scenario that takes us either to a multiverse, to religion and superstition, or—well—to nothing at all. On the plus side, the big freeze is so many trillions of years into the ultra-distant future that it's hard to envisage. If we want to put a number to this many years ahead, we could use the googol, a name coined by a nine-year-old American boy, Milton

Sirotta, in the year 1920, during a weekend walk with his uncle, the mathematician Edward Kasner. A googol is the unimaginably large number that corresponds to 10 to the power 100. But perhaps we would be spending our energies more productively by considering the much closer future in which we're planet-hoppers.

Alien Endings

We humans may not be the only planet-hoppers of the future. Let's assume that the home planet of those Andromedan scientists whom we met in Chapter 1, and the star that it orbits, are of broadly similar ages to the Earth and Sun. If that's the case, then they'll be looking for a new home planet at around the same time as we will. And since their civilization is more advanced than ours, then unless we somehow catch up with them, they'll be better than us at finding new homes. We may even be competing with them because, when our respective suns die, they'll no longer be Andromedans. Both they and we will be Milkomedans; and how close we'll be to each other in Milkomeda is an imponderable.

Now recall that the ballpark calculations we did in the opening chapter suggest there's not just one inhabited planet per galaxy but lots of them. If so, then the number of planet-hopping civilizations of the future might be large. In such a scenario, how will they interact? At the positive end of the spectrum of possibilities, the future civilizations involved, including us, may cooperate and develop a set of rules for the peaceful prioritizing of rights to colonize.

At the negative end of the spectrum is, of course, conflict. Rather than peaceful negotiation and rules, it may simply be a case of whoever is strongest, or most advanced, taking over. Indeed, such a conflict scenario need not be as distant as the death of the Sun. There may be a species of intelligent aliens "out there" whose home star is about to die. These aliens may already be eyeing possible planets to colonize before their own becomes uninhabitable. They may arrive here next year and exterminate us, especially if they are so much more advanced that they view us in the same way as we view chimps—creatures whom we have come close to exterminating ourselves.

Which is the more likely scenario—negotiation or conflict? We can't answer this question, especially as we don't know the nature of the aliens concerned. But we can look to the history of encounters between civilizations here on Earth for clues. Of particular relevance are encounters in which one civilization is more technologically advanced than the other.

An obvious example is the colonization of North America by Europeans a few hundred years ago. Before their arrival, the continent was already inhabited by humans. These inhabitants were thus of the same species as the colonizers, though they were different in many ways. They looked different. They had different societies and customs. They lived very different lives. Most of the colonizers saw not what we today would call interesting diversity and multiethnicity but rather inferiority and threat.

So, who came out as winners of the ensuing conflict—the incoming Europeans or the indigenous Americans? We all know the answer to this question. Let's hope that the outcome of humanity's future encounters with aliens who are more advanced than us is not of the same general sort.

But all this speculation about the *distant* future may be completely unjustified. Humanity may not have a distant future. Like other species of animals here on Earth, our likely future may be imminent extinction rather than planet-hopping. Let's now examine that possibility.

Chapter Twenty

Extinction and How to Avoid It

Run-of-the-Mill Extinction

The dinosaurs were famously sent to their doom by an asteroid, but most extinction is a more mundane matter. Every million years, many species of animals become extinct without the intervention of an object from space. Not even a dramatic geological event, like a volcanic eruption, need be involved. The number of individual animals belonging to the species concerned may simply dwindle, eventually to zero. An exact cause may be hard to discern. It may be that the predators of the dying species evolved faster than it did. Or perhaps the climate changed too fast for it to keep up with, evolutionarily speaking. Or perhaps there wasn't even a single cause of its extinction but rather a combination of many factors. This is how most species perish—not through a single catastrophe but through a failure to evolve fast enough to survive a period when many components of the environment are changing. And all periods of Earth history are periods of change rather than stasis. The question is not whether there will be change but how much.

This kind of inconspicuous trickle of species loss is called *background* extinction. But the choice of name is not because the failing species can be thought of as literally shrinking into the background of a landscape in which it is a smaller and smaller player. To see why this name is used, we need to think in terms of a timeline of extinction—one that extends over about 500 million years. If we use the fossil record to estimate the number of extinctions of animal species that take place every 10 million years over that great stretch of time, we'll have a series of 50 such estimates. Most of these estimated extinction rates will be similar, and you might say low—at least compared to the few that stand out from the crowd. The small number of estimates that do stand out—statisticians call them outliers—correspond to what biologists call *mass extinctions,* including the one in which the dinosaurs perished.

The number of mass extinctions in the history of the Earth so far is debatable. Depending upon whom you listen to, the number is anywhere between none and about 10. A figure of six is quite common, but don't trust it. A graph of extinction rate against time is a zigzag pattern of ups and downs. How big any particular *up* has to be for it to deserve the title "mass extinction" is a moot point. There's no clear threshold upon which everyone would agree. In the next two sections, we'll have a look at two events in the Earth's history that almost all biologists would regard as mass extinctions—the "great dying" and the death of the dinosaurs.

The Great Dying

The "great dying" was the biggest mass extinction of all in the history of the world, and it predated the dinosaurs in every sense. It occurred before the more famous mass extinction that killed them off; in fact, it occurred before the first dinosaur set foot on the Earth—so their group was not at risk in the great dying because it did not yet exist.

The great dying happened about 250 million years ago. What caused it is less certain than in the later case of the dinosaur extinction. There may have been several interacting causes, including another asteroid impact. There's a huge literature on the great dying, ranging from very detailed scholarly articles to very rough-and-ready journalistic accounts. We need to be wary of figures purported to be extinction rates, especially in the latter kind of publication. This is an appropriate place to recall the saying "There are three kinds of lies: lies, damned lies, and statistics." Even something as simple as a percentage can be misleading, either because the authors concerned were careless or because (even worse) they were going for maximum impact rather than accuracy. And another problem arises because sometimes different kinds of percentages are being inappropriately compared with each other.

For example, it's easy to find claims that up to 99 percent of marine animals, of all animals, or even of all life became extinct in the great dying. But what does this mean, especially when other, more moderate figures can also be found—for example a "mere" 50 percent of life going extinct? Part of the discrepancy is simply due to exaggeration—the figure of 99 percent should really be more like 95 percent. But most of it is due to something else. Not all estimates of extinction rate use *species* as their unit of the diversity of animal life. Species are the best units to use but also the hardest, because it's often difficult to distinguish whether a group of poorly preserved fossils represents just a single species or several closely related ones. A way to avoid this problem is to use the families into which species are grouped, rather than the species themselves.

An example will show the enormous effect of this apparently minor difference in the way of measuring the extinction rate. Suppose we're interested in those famous inhabitants of ancient oceans, the trilobites. Until the great dying, these were what we might call the ultimate survival machines. There were huge numbers of them, and they persisted for more than 270 million years, surviving earlier mass extinction events when many of their contemporaries died.

They did finally meet their demise in the great dying, so during that event their extinction rate was, regrettably, 100 percent.

Consider a very specific point in time about 250 million years ago, right in the middle of this mass extinction. Suppose at that point in time two families of trilobites, each of which originally had ten species, have suffered badly; one has a single species left, the other none. The result: a family-level extinction rate of 50 percent (1 family left out of 2), and a species-level extinction rate of 95 percent (1 species left out of 20). This nicely illustrates how apparently simple things like percentages have to be handled carefully so that they don't end up becoming those proverbial damned lies.

Now that we're better equipped to interpret claimed extinction rates in a critical way, let's see just how great the great dying was. It was very great indeed—it came close to wiping life from the face of the Earth completely. We are here because the great dying killed off "only" about 95 percent of species rather than 100 percent. Notice that I've used the species-level figure because, in the end, despite all its difficulties, this is the most *real* way to measure extinction. Species are real biological units, the borders between them being defined by lack of ability to interbreed, whereas families are mental conveniences—helpful categories to enable us to organize our view of biological diversity, but without any clear definition.

What we see, then, is that this mass extinction, the one that came closest to terminating life on Earth during its 4-billion-year history to date, happened at a clearly known point in time (250 million years ago) and killed an approximately known percentage of species (95 percent), but it happened for reasons that are far from clear. We can measure the great dying, but we can't satisfactorily explain it. There are ideas and hypotheses, but not a clear overall picture.

Of Asteroids and Dinosaurs

In contrast, the cause of the later mass extinction that killed off the dinosaurs is now well established. The battle between asteroid and volcano camps has given way to a consensus in favor of the asteroid. We know where it fell, the Yucatán Peninsula of Mexico; we know how large the crater it left is, more than 160 kilometers (100 miles) wide; and we can work out roughly how large the asteroid was, similar to a detached Mount Everest. We find evidence that its effects were worldwide rather than local from the existence of a layer of the element iridium in rocks that were formed at the time of the impact, about 65 million years ago. This element is very rare on Earth but much commoner in asteroids.

Let's picture the beginning of the end of the dinosaurs—an event that took place far from Earth in the asteroid belt. What follows is a story that's hypothetical but largely based on fact.

Somewhere in the asteroid belt between the planets Mars and Jupiter, in which more than a million large objects are orbiting the Sun, two collided with each other. The larger one was relatively unscathed by the collision, but the smaller one—still very large—was knocked out of its orbit. It was kicked out of the asteroid belt altogether on a course that was unpredictable, but chance brought it in our direction. As it neared the Earth, its trajectory was influenced by our gravity, without which its route might have been a near miss. But given the deflection caused by Earth's gravitational field, it didn't miss us, it hit. And the consequences were catastrophic.

The asteroid broke up; bits of it were scattered everywhere, leaving their iridium traces; and material from both the asteroid and the Earth was thrown up into the atmosphere, almost eliminating the passage of sunlight to the surface of our planet. Consequently, most plant life died, followed by most animal life, including the dinosaurs. The overall species-level extinction rate may have been about 80 percent—not as high as the 95 percent of the great dying, but catastrophic all the same.

Although the real version probably wasn't much different from the above fictional account / hypothesis, there are some strange things about this mass extinction that we still can't explain. For example, why did mammals survive while the dinosaurs did not? Some say it was because the mammals of the time were very small. But why should small size be advantageous? And why did lots of invertebrates much smaller than the mammals perish? Why did some groups of reptiles not go extinct? Why did the crocodiles, which were larger than small dinosaurs, survive? Some say that it was because they were aquatic. But why should living in water help? And if it does, why did so many marine invertebrates perish? These unanswered questions should urge us to be cautious about our understanding of the precise sequence of events from impact to extinctions. There is still much that we do not understand.

Multiple Dangers

All species of life of Earth, including humans, are subject to a variety of threats to their future existence. The main categories of threats are astronomical, geological, climatic, and biological, the last of these categories including disease. But although humans are in the same boat as all other animals in this respect, we're special too, in relation to *causing* extinctions.

First, we ourselves are a significant cause of extinctions simply as a result of our day-to-day activities, which result, among other things, in climate change. Some scientists think that the Earth is currently entering a new mass-extinction event, of which we are the primary cause. No other species on its own has such major global effects that threaten so many other creatures. Our stewardship of the planet is crucial not just to our own survival but to the survival of life in general. Ephemeral headlines that appear in the news from time to time on attempts to reach an international agreement to cut greenhouse gases really are crucially important, despite the fact that they, like all other news headlines, give way to more recently breaking stories soon after they appear. In the battle for people's attention, the recent and urgent tend to win over the long-term important. If we are to survive as a species, we must not lose sight of the latter.

Second, from time to time, in addition to the negative effects of our everyday activities, we humans have wars. These kill significant numbers of our own species and of other species too. In the First World War, more than 15 million people died—and to what end? we might ask. In the Second World War, the deadliest of all human conflicts so far, the death toll was approximately four times that of the First. Although these numbers are huge, and every individual death was a tragedy, they are small in percentage terms—"only" about 2 percent of the human population died in the Second World War, as opposed to the 100 percent death that constitutes extinction. But, regrettably, future wars have the potential to achieve the latter percentage with ease.

Our unique ability to wipe ourselves out with nuclear weapons might extend to wiping out most or all species on the planet in addition to our own. In other words, something that is unique to us—global-scale armed warfare—may also make us a unique cause of extinction across the board. Not only did millions of humans die in the First World War, so did millions of other animals, ranging from the horses that were used in conflict to the small soil animals whose humble abodes were blasted into the sky by bombs. As with humans, these animal death tolls were large in numbers but small as percentages, so they weren't causes of extinctions. But their future equivalents might be.

Avoiding Human Extinction?

With apologies to the rest of the animal kingdom, I'd like now to focus entirely on our own species and the issue of how we can survive into the distant future so that our descendants can witness the merger of the Milky Way and Andromeda in about 4 billion years' time. We should perhaps begin this discussion by reminding ourselves that such long-term survival is extremely unlikely. Most successful species of the past have lasted just a few million years. A small number, the so-called living fossils, have lasted longer, but probably no single species has lived for more than 100 million years. So 4 *billion* years would be amazing.

At this point we need to understand something called pseudo-extinction. This is what happens when, instead of genuine extinction, a species ceases to exist but leaves descendant species that are a bit like it. For example, triceratops, dinosaurs easily recognized from the three formidable horns after which they're named, became genuinely extinct 65 million years ago. As far as we know, not a single living animal today has a triceratops as an ancestor. In contrast, the group of dinosaurs known as Maniraptora, which included the velociraptors made famous by the movie *Jurassic Park,* only experienced a pseudo-extinction because today's birds all belong to this group, having descended from a maniraptoran ancestor of the Jurassic period.

When considering human survival or extinction over such long periods of time as billions of years, we should acknowledge that one form of survival—indeed, the most likely one—is that the "humans" of the distant future will be descendants of, not members of, *Homo sapiens,* which would mean that our species suffered a pseudo-extinction. Such a fate would be fine with me, and I hope with you too. Our species has progressed far since the times of our handy person ancestors, the species *Homo habilis,* which we looked at in Chapter 2, but we're far from perfect. If a descendant species (say,

Homo astronomus) were to represent us in 4 billion years' time, that would be an evolutionary success, not a failure. The key question now becomes: what are the most likely causes of our extinction in the meantime, especially in the near future, and how can we try to avoid them?

The remainder of this chapter thus has a tough task—the taking on of which could easily be expanded into an entire book. I did indeed consider writing a whole second book about this subject, but decided against it: a book that's *entirely* about future possibilities would be hard to grapple with, and anyhow there are other authors better qualified than I am to undertake such a task. But what can be said on the subject of avoiding human extinction in just a few pages, especially if it's not merely to be a statement of the obvious? Let's see.

We'll now look at two possible causes of human extinction within the next few thousand years. They're not the only ones—recurrence of an asteroid impact is another possibility—but in my view they're the most likely.

Going Viral

The first of these causes is disease, and specifically those diseases that are caused by microbes—loosely defined to include both bacteria and viruses. Inherited diseases can have appalling consequences for individuals, but they're not a threat to our species. Diseases caused by animals, such as schistosomiasis, the disease agent here being a parasitic worm, can also have very adverse effects on the people suffering from them, but they are unlikely to cause human extinction. In contrast, diseases caused by microbes are a potential threat to the continued survival of humans on planet Earth.

The reasons for this assertion are as follows. The shorter the life cycle of an organism, the faster it can evolve. And when a disease agent has a very short life cycle and its host a very long one, the discrepancy between the two is most likely to result in an evolutionary advantage to the former. Also, the changing fates of host species in terms of their population numbers will be a major factor driving the evolution of the disease agent. Once upon a time, a few million years ago, the population numbers of protochimpanzees and protohumans were probably about equal. Now there are less than half a million chimps and more than 7 *billion* humans—so there are about 10,000 humans for every one chimp. If you were a bacterium currently specializing in chimps, what evolutionary trajectory would you choose?

When antibiotics were discovered, in the early twentieth century, they were a boon to medical science. They cured many bacterial diseases. And to some extent they still do. But, given the ability of bacteria to evolve very rapidly, we now live in the age of so-called superbugs, such as MRSA (a label for tough strains of the *Staphylococcus aureus* bacterium), which are resistant not just to one antibiotic but to lots of them. The superbacteria of the future will be even more problematic, and may constitute an extinction threat. What tactics should we take in order to avoid this fate? The short answer is to dramatically increase the funding for biomedical research.

Superbacteria are a real threat, but the threat posed by viruses is even greater. Contrary to what some people think, most disease viruses are unaffected, or only negligibly affected, by antibiotics. If you have influenza, which is caused by a virus, and you take antibiotics, you won't be cured by them (though you probably will be helping bacteria to evolve antibiotic resistance). Viruses are harder to deal with than bacteria; vaccines work well in some cases, but in other cases we're still struggling to find an effective one.

Think of the serious viral outbreaks that have affected humans over recent times. Some of those that most readily spring to mind are smallpox, Ebola, and AIDS. Smallpox has been eradicated (we think) through effective vaccination campaigns. Ebola, regrettably, has not; at the time of writing there is no effective vaccine, but there is much ongoing research to try to find one.

AIDS is worthy of particular attention because of the part of the body that the virus concerned attacks, and because future versions of this type of virus may pose an even greater threat to humanity than the current versions. All disease viruses target some parts of the body more than others. But in the case of the human immunodeficiency virus (HIV), which causes AIDS, it primarily targets cells of the immune system—the very system whose job it is to protect us from infections.

This is a bit like an enemy state having developed a weapon that only affects our military personnel: something that causes soldiers to die so that our enemy can invade our country and enslave our civilian population. The parallel isn't perfect, because a civilian can become a soldier, whereas a brain cell cannot become a white blood cell of the immune system. Nevertheless, the analogy helps to illustrate the dangers to an organism of hostile agents specifically targeted at that entity's defense system. As with Ebola, there are no effective AIDS vaccines as yet. There are some treatments that are partially effective, such as the drug AZT; also, combinations of drugs can be more effective than single ones. But not only are these drugs inadequate right now, they may become even less adequate in the future. Just as with bacteria evolving resistance to antibiotics, viruses can evolve resistance to the various agents used against them.

The answer, again, is to boost biomedical research. However, greater funding is just a start. Money makes medical research possible but doesn't guarantee its success. The main factor in such success is the ingenuity of the researchers. Our foremost distinguishing feature as a species, our large brain, may also be our savior.

Nuclear Oblivion

Even without antiviral agents, a small percentage of people often turn out to be naturally immune to any particular virus. Seven million survivors out of 7 billion, a tiny 0.1 percent human survival rate, could be enough to avoid the extinction of humanity from a viral pandemic. But no one is immune to high doses of ionizing radiation, such as those that result from the explosion of a nuclear weapon.

The year 1945 was a historic one for both good and bad reasons. It is most widely remembered as the end of the Second World War, the deadliest war in history. But the final stages of the war involved the first, and so far only, use of nuclear weapons in human conflict. Approximately 50 million people died in the war overall, including about 150,000 victims of the bombings of Hiroshima and Nagasaki.

In the 70 or so years since, no nuclear weapons have been used in warfare, despite the fact that there have been many wars and there exist in the world today many thousands of these weapons, possessed in various numbers by nine countries. The vast majority are owned by the United States and Russia, but significant numbers exist in other countries, specifically China, North Korea, India, Pakistan, Israel, Britain, and France. We humans now have at our disposal the means of making planet Earth virtually uninhabitable.

Can we as individuals, nations, or a species do anything to avoid, or at least to reduce the probability of, a self-inflicted nuclear extinction? Unlike the threat from superbugs, money is not the answer here. Rather, the answer lies in the future upbringing and education of the people who will take control of, and eventually either abolish or use, these fearsome weapons. If the weapons are abolished, the money that would otherwise have been used for their maintenance and replacement could in theory be given to biomedical research, thus killing the proverbial two birds with one stone. It's a shame that reality isn't so simple. Anyhow, we'll move on, in Chapter 21, to this issue of upbringing and education, which is so crucial to our future survival. In particular, we'll address the question of what influences the likelihood that humans grow up to be enlightened or intolerant adults. But as a prelude to this sharp focus on ourselves we'll expand outward one last time . . .

Chapter Twenty-One

From Embryo to Enlightenment

Alien Development

We saw earlier that Darwinian natural selection is likely to apply to alien life forms, because only three attributes are required for it to take place: reproduction, variation, and inheritance. Whether the nature of alien inheritance would be similar to ours—bound by Mendel's laws—is less certain. In terms of the structure of alien organisms, I suspect that, like almost all life on Earth, they will turn out to be composed of cells. Such a view can be termed cellular chauvinism; and this, like carbon chauvinism, is a view to be held only provisionally. Nevertheless, it's worth noting how little life on Earth has been able to achieve without cells. The only known forms of noncellular organism on our planet are viruses (smaller than cells) and creatures such as slime molds (bigger than cells but undivided). Viruses are only able to exist as parasites of other life forms, and none of the acellular slime molds have managed to evolve complex organ systems comparable to those of animals and plants.

If alien life is indeed cellular, an interesting question arises: do multicellular aliens have life cycles that require them to go back to a unicellular stage at the start of every generation, as we humans do?

In this context, comparative information on some of the million or so known species of animals on planet Earth is interesting, but before we examine it we should cast an eye on the plant kingdom—the second biggest terrestrial experiment in multicellularity.

Although flowering plants typically have a unicellular stage in their life cycle, with an egg that gets fertilized by pollen, they can reproduce asexually, in which case a return to a unicellular stage is not required. I have grown many plants from cuttings taken from "parental" plants of the same species. The ability of a piece of tissue from one plant to become another complete plant is very widespread.

However, this phenomenon is comparatively rare in animals. Half a worm can become a new worm; but half a monkey is, regrettably, a developmental dead end. In general, "higher animals" (however ill-defined that term may be) cannot regenerate from a body part. Admittedly, some vertebrates have the ability to do what might be described as the opposite: regenerate a body part from a body. For example, newts can regenerate severed legs. But a newt leg cannot regenerate the rest of the newt. And no vertebrate can regenerate a head.

Perhaps in the case of *intelligent* alien life a similar "law" applies. Perhaps they too must go back to a unicellular stage once in every life cycle. If this is so, then not only are there alien eggs but alien embryos too, because to produce an adult multicellular organism from a fertilized egg (or equivalent) requires a process of development.

So I've nailed my colors to the mast. I'm a firm believer that many of the "laws" of biology, however exception-prone they may be compared with the laws of physics, will apply to life on other planets, perhaps right across the universe. But we can only go so far with this line of thought until we have some evidence. For now, it's prudent to desist from further speculation. To consider whether alien embryos have mulberry and sausage stages in their life cycles, for example, would be a step too far into the void. And as for alien post-embryonic *mental* development, there are far too many unknowns for us to come up with detailed hypotheses. However, in terms of the broad-brush picture, I wonder whether curiosity may play an important role in alien mental development, just as it does in the mental development of humans.

Curiosity and Its Quashing

Just now, as I'm writing this, a robotic vehicle called *Curiosity* is taking samples of material from the surface of Mars, looking for signs of past or present microbial life. It may still be engaged in this task as you read these words. It began its work in 2012 when it landed in the Gale Crater, a vast impact depression on the Martian surface, the center of which features a mountain, Aeolis Mons, that is higher than Mont Blanc.

This Martian rover couldn't have been better named. Our species is naturally curious. Of course, so are many other animals, but curiosity seems to reach a peak in humankind. Young children are full of questions: "What's this? How does it work? Why is it here?" And so on. You've probably had the same conversation with a child as I have—the one that's a series of iterations of "Why?" Usually the first "because" is easy, the second one harder, and soon we reach a "Why?" that's difficult or even impossible to answer.

Adults who try to give children answers to questions, regardless of whether they begin with what, how, or why, have a huge responsibility, whether they are parents, teachers, or others. When we interact with children in this way, we are not just imparting information; we are also helping to determine their ways of thinking. We are contributing to the direction of their mental development and hence helping to mold their minds. We can be encouraging reason or rubbish, logic or superstition, sense or nonsense.

Example: A child asks an adult, "Why does the Sun move across the sky?" Bored adult response: "Look, it just does. Stop bothering me." Misinformed adult response: "Because it's orbiting the Earth."

Informed adult response: "Actually, things aren't always what they seem. Although it looks like the Sun is going around our sky, and around us, the truth is that we're orbiting it, and spinning on our axis at the same time. The Sun appearing to go around us is an illusion caused by the Earth spinning."

Of course, age matters. The wording used by the informed adult in this case would only be appropriate for certain ages of children. For *very* young children, different words have to be used, because the right answer given in the wrong form is almost as useless as the wrong answer. So again we see the burden of responsibility on the part of the adult who is doing the explaining.

Although the above example illustrates the difference between three types of adults, there's a fourth type we need to consider: religious fundamentalists, be they Christian, Jewish, Muslim, Hindu, or of another faith. In my view, such people pose the greatest threat of all to the inquiring young minds of the twenty-first century. They continue to crush the natural curiosity of children, just as they've done in the past, under a mass of supposed facts for which they have no evidence, often simultaneously poisoning the children's minds with hatred against their fellow humans who do not believe these particular "facts."

Liberal religion that is freethinking and tolerant of the views of others is not a problem—indeed, countless acts of kindness, charity, and genuine education (as opposed to indoctrination) have been carried out in its name. But unfortunately, thus far through human history, most organized religion has not been of this enlightened sort.

As we saw earlier, the Italian philosopher Giordano Bruno was burned at the stake in Rome in 1600 by the Catholic Church. The reason for this appalling murder was that Bruno dared to believe things that they didn't, including that stars were distant suns, which might be hosts to planetary systems and life. This is regrettably just one incident out of many where a hard-line religion has murdered people just for thinking for themselves and refusing to accept its dogmatic beliefs.

In 2015, the noted Syrian archaeologist Khaled al-Asaad, an 81-year-old scholar, was beheaded in the city of Palmyra by the terrorist organization that calls itself the Islamic State (IS). The reason for his murder, which was preceded by torture, was his refusal to reveal the locations of ancient treasures associated with polytheistic religions. Two temples dedicated to gods of these religions were blown up by IS at about the same time. The destruction of these important ruins in a UNESCO World Heritage Site is an appalling crime in itself. But it pales into insignificance when compared to IS's crimes against humanity, including the murder of innocent civilians like al-Asaad, whose decapitated body was hung up in the center of Palmyra as a gruesome warning to others not to go against the violent fundamentalism of IS.

The killings of Bruno and al-Asaad were both carried out, supposedly, in the name of a good God. These two murders, four centuries apart, remind us that supreme evil is not the sole preserve of any one religion or political movement. Rather, it's something that most religions are capable of spawning. Wherever such evil breaks out, whether in seventeenth-century Rome or present-day Syria, its distinguishing features include extreme intolerance to freedom of thought and the violent imposition of dogmatic authority. Usually, the ultimate source of such authority is deemed to be a book.

Books, Enlightenment, and Intolerance

This book began with a quote from that enlightened seeker after truth and understanding, Thomas Henry Huxley. It's now time for another of his many wise sayings. Huxley once stated (1896, 226) that "the ultimate court of appeal is observation and experiment, and not authority." What he meant was that in order to judge for ourselves whether some claimed "fact" is true, we should not look either to the verbal pronouncements of authoritative figures or to the written pronouncements of authoritative books for an answer. Rather, we should try to find out for ourselves.

Laudable though this view is, it has its limitations. If an expert on molluscs tells me it's a fact that all pond snails breathe water rather than air, I can test this claim by observation. I can watch a snail in my pond come up from the bottom and turn its respiratory opening to the water surface, exchanging gases with the air above the pond and then returning to its usual aquatic existence. It doesn't seem logical that an aquatic animal should have to do this, any more than it would seem logical for humans to breathe water, thus having to keep immersing our heads in ponds in order to survive, but it's true all the same. And it's no more illogical than a whale having to come to the surface of the sea to breathe air, when its fishy cousins don't. Evolution is a messy business. Anyhow, my own personal observations enable me to realize in this case that the dogmatic "authority" on molluscs was wrong.

On the other hand, if a dogmatic cleric comes up to me and tells me that there is a divine being who insists that I should believe in "Him," and not only that but I should also believe that He has a particular name (different in one religion from another), it's hard for me to disprove the cleric concerned. However, if he tells me that the basis of his claim is a particular book (again variable between religions), I can at least read the book critically, think about the issue independently, and decide for myself—though that's undoubtedly not the kind of reading that the cleric would have wished me to do.

Near the start of the book of Genesis, in the Christian and Jewish Bibles, we are told that planet Earth had gotten to an advanced state of creation by the end of day three, with the continents having been separated from the oceans, and many types of plants having appeared. The Sun and the other stars were not created until day four. Two thousand years ago, it might have been reasonable to believe this account, but today most people recognize it as false. Indeed, many Christians and Jews recognize it as false, though some might prefer to use the descriptor "figurative." However, many still claim the literal truth of the Bible, and not just in the Deep South of the United States. Here's an extract from the website of a church in my hometown, Lisburn: *We believe the Bible, as originally given, to be without error, the fully inspired and infallible Word of God.*

It's no criticism of the authors of Genesis that they got things wrong. They were writing a long time ago, in an era without advanced science and technology. They had no telescopes through which to look at the sky, nor any means of dating terrestrial rocks. They were clearly thinking about the big questions at a time when most people spent most of their daily lives just trying to survive.

One part of a book being wrong does not mean that the whole book is wrong. Indeed, all books have good bits and bad bits. So we don't dismiss the Bible as a whole; it contains many important ideas that have stood the test of time much better than those expressed in Genesis on the origin of the Earth, the Sun, and the stars.

Other holy books are much the same. Some parts of their content have been superseded by advancing human knowledge; other parts have not. The biblical commandment not to kill our fellow humans, versions of which are found in many other holy books, is as relevant today as it was millennia ago. Those who burned Giordano Bruno to death apparently chose to ignore it, which is somewhat ironic.

Let's shift our attention from holy books to science books. I have a copy of a textbook on evolution written in the 1940s. Some of its content is still thought to be correct, but some of it is known to be wrong. Old textbooks on genetics I've simply thrown away, because, given the advances in that subject, the books concerned had become useless. A present-day biologist who chose to cling to a textbook on genetics from the early twentieth century because it was the first such book and was infallible would be regarded by colleagues as crazy.

In science, we read books and other written material (websites, journal articles, and so on), but we do so critically, trying to understand as we go along, and questioning anything that doesn't seem to make sense or that is badly explained. Books are a means to an end—progress in human understanding. If anything is holy in this context, it is the quest for understanding, and the freedom of thought that makes understanding possible.

How are we to reconcile the views of books as permanent and infallible on the one hand and as transient and imperfect, on the other? The American evolutionary biologist Stephen Jay Gould suggested, in his 2002 book *Rocks of Ages* (5), that science and religion are two "Non-Overlapping Magisteria." Put simply, he claimed that science deals with facts but religion deals with the question of how we should live our lives, and the question of what happens to us when we die. There's clearly a distinction between science and religion of the approximate sort that Gould suggests. However, it's messy rather than clear, so the *non-overlapping* is unjustified. The Bible gives us both "facts" and advice on how to live. So do other holy books. The main problem with the holy books is not so much that some of their facts are wrong, but rather that religious fundamentalists set these books aside from all others and regard them as being unquestionably true—sources of wisdom that simply must be believed in and adhered to in all respects.

The key word here is *unquestionably*. If we allow anything to be beyond the realm of questioning, we give up part of our humanity. Our mental development is thwarted. We do not achieve enlightenment in any sense of the word. There should be no areas of human thought that are immune to the question "Why?" regardless of whether it comes from an innocent three-year-old child, a teenage student, or an octogenarian. If there is anything a good God would hate, I suspect it would be the renouncement of our curiosity, along with our ability to question things without restriction.

So, books of all kinds are to be read; but they are to be read critically, not unquestioningly, and I see no reason why religious books should be exceptions. On finishing a particular book, we might end up believing the main things the author says. Or we might not. In either case, we should use our powers of reasoning, along with further reading from other sources, and personal observations where possible, to assess whatever points the author is making. We should never believe things uncritically just because they're written down. After all, words on a page or a screen are no different from the words that we speak. And in my opinion, none of our words are dictated by God; all human words have human sources. But many people have a contrary opinion; and therein lies the problem.

At the other extreme from revering a book as being holy is burning it, yet these opposites are closely connected with each other. And they're both linked to a dogmatic mind-set. If one book is holy and infallible, then another that appears to contradict it is a candidate for burning. Regrettably, history is littered with public actions of this kind. They haven't all been primarily to do with religion—Nazi book-burning in the 1930s is a case in point. But many of them have had a religious motivation.

In the year 650 or thereabouts, versions of the Quran that didn't correspond to the "true" one were ordered to be burned by the third caliph, Uthman ibn Affan. In 1520 one of the founders of Protestantism, Martin Luther, had a book by Angelo Carletti, the *Summa Angelica,* publicly burned in the town square of Wittenberg because of its ultra-Catholic stance. In 1624 the Pope ordered the burning of copies of Martin Luther's German translation of the Bible. In the 1720s rabbis in Italy ordered the burning of books by Moshe Chaim Luzzatto, whose mystical ideas were considered to be heretical to the mainstream Judaism of the time. And so it goes on.

Burning your opponents' books is better than burning their authors, but these two symptoms of intolerance are closely linked. The Nazis burned books by Jewish authors; later, they notoriously killed Jews in their millions, in the worst genocide of recorded history. Protestant-Catholic enmity has also extended from book burning to mass murder. In the St. Bartholomew's Day massacre in 1572, thousands of Huguenots (French Protestants) were killed by Catholics in Paris, one of many instances of their widespread persecution. And there have been many persecutions and murders of Catholics by Protestants too. Enmities between religions, or even between branches of supposedly the same religion, resulting in the burning of books and the killing of people, are all too readily found throughout history. And such behavior is not over yet as we see, especially in the Middle East, today.

Living Together, Living Apart

So, one newborn child develops into a rational, empathetic adult, respecting human diversity; another develops into a hate-filled fundamentalist who finds it acceptable to murder his or her fellow humans because they have different beliefs. The process of mental development that culminates in either of these fates, or an intermediate one, is complex, but it does have a pattern in time. We are most easily influenced when we're young. Our "influenceability" decreases with age. The founder of the Jesuits, Ignatius Loyola, is supposed to have said, "Give me a child to the age of seven and I will give you the man." This is an apocryphal quotation, but regardless of its source, it carries both a truth and a falsehood.

The falsehood arises from the specification of a particular age threshold below which we are mentally malleable, above it not. Our decreasing mental malleability has a time trend, and during some periods of time the decrease is faster than others, but there isn't a specific cut-off. Children older than seven, but not by much, have had their mental development switched from a trajectory toward reason to a very different one toward violence—the child soldiers of recent wars in Africa and of the killing fields in Cambodia being prime examples. And the radicalization of teenage Muslims via the Internet is another dark manifestation of our post-age-seven ability to be derailed from a mental path toward enlightenment and redirected onto one toward intolerance.

Although *reading*—books, websites, newspapers—provides much information and helps to influence our mental development, it's not the only such thing and perhaps not even the most important one. The young child who repeatedly asks "Why?" has been influenced more by interactions with other people than by books. Our early development is influenced by those around us—parents, teachers, friends. And lack of exposure to a *diversity* of people and cultures can be a limiting factor in the development of our enlightenment.

As a child I grew up in the northern part of the island of Ireland—in a town that is within both the historic Irish province of Ulster and the present-day state of Northern Ireland, the part of the island that's still politically linked to Britain. Although about a third of my hometown's residents were Catholics, I never had any Catholic friends. My own upbringing was Protestant, though I no longer claim such a religious allegiance, or indeed any—as noted earlier, I'm an agnostic, like my scientific hero Thomas Henry Huxley, who coined the term. I lived in a street with about 30 houses, all of which were lived in by Protestant families. There were many such streets, and many others that were exclusively Catholic. I went to a school at which there were no Catholics. My Catholic counterparts went to a school where there were no Protestants.

This absurd situation had historical roots. I belonged to the Ulster-Scots ethnic group, descended from the "planters" who came to Ireland from Britain in the 1600s. My fellow Catholic townspeople belonged to the Celtic-Irish ethnic group—descended from the Celts who came to Ireland from mainland Europe much longer ago. I grew up in a system of voluntary apartheid, though it took me a long time to realize that fact.

Many areas of Northern Ireland were the same as mine in terms of voluntary segregation both in residence and in education. The Troubles, the sectarian violence that persisted from the late 1960s to the late 1990s, and resulted in the murder of more than 3,000 people and the maiming of many more, might never have happened if the two communities had lived and been educated together. Maybe that's wishful thinking, but maybe not. Some of the root causes of the Troubles might not have persisted until the violence started if the two communities had been better integrated.

For example, from talking to Catholic friends, Protestants might have realized that the Northern Ireland parliament as it existed in the 1960s was both a democracy and a dictatorship: the former because members were elected, the latter because most people voted on the basis of ethnic group membership, with the result that the same party was always in power. If we had all been able to appreciate the full significance of this in the 1960s, we might have voluntarily switched to the sort of power-sharing Assembly that now exists, albeit precariously, in Northern Ireland, rather than getting there via thirty years of violence, death, and destruction. And the best way to prevent a recurrence of violence lies in desegregation. The Integrated Education movement is an excellent step in the right direction, but it's only a start; much remains to be done.

Northern Ireland is just a tiny part of the world. But the lessons we've learned there have wide applicability across all nations. The most important thing is that as children grow up they should meet and mix with others from all the other religions, races, and ethnic groups that exist in their country—and if possible beyond. *We all need to learn to see each human being as an individual, not as a representative of a group. And the younger we learn to see people this way, the better.*

It's easy to point out and discuss such problems, much harder, of course, to solve them. But solve them we must if we're to survive as a species. The single biggest threat to humanity may be the coming together, in the near future, of religious fundamentalism and nuclear weapons. The policy of mutually assured destruction, which underlay the nuclear weapons standoff between the United States and the Soviet Union until the collapse of the latter in 1991, actually worked, crazy as it was; and perhaps it's still working between the United States and post-Soviet Russia. But it won't work if religious fundamentalists become nuclear-armed. There's little point in threatening to kill a would-be suicide bomber.

The Silence of Space

It's hardly surprising that there's no noise in space. Sound waves need a medium, such as air, through which to travel; they can't travel through a vacuum. But lack of sound is not the kind of silence I'm referring to here. Rather, it's the lack of radio messages. So far, despite about half a century of SETI researchers listening for radio signals from alien civilizations, we've heard precisely nothing. This lack of incoming messages seems odd, especially when considered against the background of the very high probability of alien life, which we discussed earlier. Perhaps more than odd: contradictory or paradoxical. Indeed, it is often referred to as the Fermi paradox, after the Italian physicist Enrico Fermi, who drew attention to it.

There are many possible explanations of the Fermi paradox; here are three of the main ones. First, there's nobody out there and there never has been. Second, aliens exist but they're so far away that their messages, even traveling at the speed of light as radio waves do, haven't reached us yet, or are too weak to detect after traveling such a great distance from their source. Third, most or all civilizations are so short-lived that they only have time to send out a few rudimentary signals before they wipe themselves out, perhaps by a combination of nuclear weapons and religious fundamentalism, as discussed above for the Earth. Which, if any, of these interpretations of a lack of incoming signals do you think is correct? Personally, I very much doubt that it's the first.

What about outgoing signals? The Italian radio pioneer Guglielmo Marconi first sent a radio signal across the Atlantic, from Ireland to

Newfoundland, in 1901. The first radio signals deliberately aimed out into space were sent by Russian scientists in the early 1960s. More powerful signals, including the famous Arecibo message, were sent out by American scientists in the mid-1970s. A signal that was sent out in the year 1967 would, at the time I'm writing this, have traveled 50 light-years. If there's an alien civilization on an exoplanet 50 light-years away, it will be receiving that signal about now. Even if it responds immediately, we won't receive the reply until the year 2067. The fraction of our Milky Way galaxy that's within a distance of 50 light-years from Earth is much less than 1 percent. And the fraction of any of the other billions of galaxies in the universe that's within 50 light-years is, of course, zero.

We're almost certainly not alone in the universe, but we may be alone in our local corner of our home galaxy. However, in at least one sense it doesn't really matter on which spatial scale we're the only intelligent life-forms. Our business as enlightened humans is to do what Huxley suggested more than a century ago in the quote with which this book started. It's such an important statement that we should reinspect it here: *The known is finite, the unknown infinite; intellectually we stand on an islet in the midst of an illimitable ocean of inexplicability. Our business in every generation is to reclaim a little more land, to add something to the extent and the solidity of our possessions.*

Our *other* business in every generation is to try, in all possible ways, to improve the educational and social systems of whatever country we live in so that more of its inhabitants become enlightened and tolerant, like Huxley himself. The more who do, and the fewer whose mental development leads them along a trajectory to intolerance and violence, the longer humanity is likely to last. It would be nice to think that when, in the future, radio signals from alien civilizations finally arrive on Earth, there will be someone here to receive them, and to reply.

BIBLIOGRAPHY

ACKNOWLEDGMENTS

INDEX

Bibliography

Modern Scientific Texts

Gilbert, S. F. 2014. *Developmental Biology.* 10th edition. Sinauer, Sunderland, Massachusetts.

Green, S. F., and M. H. Jones, eds. 2015. *An Introduction to the Sun and Stars.* Second edition. Cambridge University Press, Cambridge.

Jones, M. H., R. J. A. Lambourne, and S. Serjeant, eds. 2015. *An Introduction to Galaxies and Cosmology.* Second edition. Cambridge University Press, Cambridge.

Maynard Smith, J., and E. Szathmáry. 1999. *The Origins of Life: From the Birth of Life to the Origin of Language.* Oxford University Press, Oxford.

Minelli, A. 2009. *Perspectives in Animal Phylogeny and Evolution.* Oxford University Press, Oxford.

Rothery, D. A., I. Gilmour, and M. A. Sephton, eds. 2011. *An Introduction to Astrobiology.* Revised edition. Cambridge University Press, Cambridge.

Rothery, D. A., N. McBride, and I. Gilmour, eds. 2011. *An Introduction to the Solar System.* Revised edition. Cambridge University Press, Cambridge.

Popular Science

Cadbury, D. 2001. *The Dinosaur Hunters: A Story of Scientific Rivalry and the Discovery of the Prehistoric World.* Fourth Estate, London.

Carroll, S. B. 2005. *Endless Forms Most Beautiful: The New Science of Evo Devo and the Making of the Animal Kingdom.* Norton, New York.

Carroll, S. B. 2006. *The Making of the Fittest: DNA and the Ultimate Forensic Record of Evolution*. Norton, New York.

Cohen, J., and I. Stewart. 2002. *Evolving the Alien*. Ebury Press, London. (Later editions published under the title *What Does a Martian Look Like?*)

Dawkins, R. 1986. *The Blind Watchmaker*. Penguin, London.

Gould, S. J. 1987. *Time's Arrow, Time's Cycle: Myth and Metaphor in the Discovery of Geological Time*. Harvard University Press, Cambridge, Massachusetts.

Gould, S. J. 2002. *Rocks of Ages: Science and Religion in the Fullness of Life*. Vintage, London.

Knoll, A. H. 2015. *Life on a Young Planet: The First Three Billion Years of Evolution on Planet Earth*. New edition. Princeton University Press, Princeton, New Jersey.

Marantz Henig, R. A. 2000. *A Monk and Two Peas: The Story of Gregor Mendel and the Discovery of Genetics*. Phoenix, London.

Shubin, N. 2008. *Your Inner Fish: The Amazing Discovery of Our 375-Million-Year-Old Ancestor*. Penguin, London.

Taylor, S. R. 1998. *Destiny or Chance: Our Solar System and Its Place in the Cosmos*. Cambridge University Press, Cambridge.

Taylor, S. R. 2012. *Destiny or Chance Revisited: Planets and Their Place in the Cosmos*. Cambridge University Press, Cambridge.

Wolpert, L. 2008. *The Triumph of the Embryo*. New edition. Dover, Mineola, New York.

Historical Texts

Darwin, C. 1859. *On the Origin of Species by Means of Natural Selection, or the Preservation of Favoured Races in the Struggle for Life*. Murray, London.

De Beer, G. 1958. *Embryos and Ancestors*. Oxford University Press, Oxford.

Haeckel, E. 1896. *The Evolution of Man: A Popular Exposition of the Principal Points of Human Ontogeny and Phylogeny*. Volume I. Appleton, New York.

Hutchinson, G. E. 1965. *The Ecological Theater and the Evolutionary Play*. Yale University Press, New Haven, Connecticut.

Huxley, L. 1900. *Life and Letters of Thomas Henry Huxley*. 2 volumes. Macmillan, London.

Huxley, T. H. 1881. *The Crayfish: An Introduction to the Study of Zoology.* Third edition. Keegan Paul, London.

Huxley, T. H. 1887. On the Reception of the Origin of Species. In F. Darwin, *The Life and letters of Charles Darwin,* Volume II, John Murray, London, 1887. (Frontispiece quote, p 204.)

Huxley, T. H. 1896. Biogenesis and Abiogenesis. In *Collected Essays,* Volume 8, *Discourses: Biological and Geological.* Appleton, New York.

Owen, R. 1843. *Lectures on the Comparative Anatomy and Physiology of the Invertebrate Animals.* Longman, London.

Paley, W. 1802. *Natural Theology: Or Evidences of the Existence and Attributes of the Deity, Collected from the Appearances of Nature.* Faulder, London.

Richards, R. J. 1992. *The Meaning of Evolution: The Morphological Construction and Ideological Reconstruction of Darwin's Theory.* University of Chicago Press, Chicago.

Wallace, A. R. 1889. *Darwinism: An Exposition of the Theory of Natural Selection with Some of Its Applications.* Macmillan, London.

Fiction, Poetry, Autobiography

Adams, D. 1979. *The Hitchhiker's Guide to the Galaxy.* Serious Productions, London.

Brown, C. 1998. *My Left Foot.* Vintage, London. (Original edition 1954.)

Carroll, L. 1871. *Through the Looking-Glass, and What Alice Found There.* Macmillan, London.

Eliot, T. S. 1925. *Poems 1909–1925.* Faber and Gwyer, London.

Acknowledgments

I'm very grateful to those friends and colleagues who read draft chapters or, in some cases, the entire draft manuscript. My brother, Chris Arthur, whose background is in the humanities and creative writing, read several chapters and succeeded in making me rewrite sections that were on the wrong side of the border between opaqueness and transparency. Colin Lawton, a fellow zoologist and Irishman, read the final chapter and reassured me that it read just as well through green spectacles as through orange ones. Ariel Chipman, a fellow specialist in evo-devo, but one who started in embryology (in contrast to my own origin in evolutionary biology), cast a critical eye over the whole manuscript, and especially its embryological parts, making comments that led to considerable improvements. Among other things, he persuaded me to reduce the number of sausage metaphors from three to one, so the only sausages you will find in the book are embryonic ones. Fred Stevenson, astronomer and cosmologist, also critically read the entire manuscript, correcting a few astro-blunders and injecting along the way calculations that allowed me to state some astronomical facts more accurately than before. I was particularly appreciative of his working out that the Milky Way (excluding Earth) contains enough of the organic compound we call alcohol for not just one of us (original version) but the whole of humanity (final version) to get very drunk indeed.

This is my first book to be published by Harvard University Press. The staff there have been a pleasure to deal with in every way. Special thanks go to my editor, Janice Audet, for her enthusiasm for this somewhat unusual project right from the start, and for her exemplary economy with words—something I have tried to learn from. Whether I have succeeded in this respect—and indeed in all others—you, the reader, are in the best position to judge. Louise Robbins caught a chapter that was trying to break out of my triplet structure, and helped me to bring it back into the fold. Tim Jones worked his magic on the cover design. And Deborah Grahame-Smith, of Westchester Publishing Services, guided me through the temporary darkness of the copy-editing process and out into the clarity that lies beyond.

Finally, I would like to thank all my friends at the Kielder Observatory, hidden among the rolling hills and conifer forests of Northumberland, just a stone's throw south of the English-Scottish border, for their camaraderie. Kielder has proved to be a wonderful place to discuss astrobiological issues with professional and amateur scientists, and with the general public. It's hard to beat a chat about the mysteries of life in the universe under the spectacular canopy of the Milky Way.

Index